Eco-Care Essentials

Your Guide to a Sustainable Future

Jeffrey S. Doyal

INTRODUCTION

clearing the Way Towards Feasible Living: Exploring a Greener Future

In a world wavering near the precarious edge of environmental unevenness, the quest for practical living has arisen as a reverberating source of inspiration. An energizing cry rises above geological limits, social contrasts, and financial layers. The desperation to orchestrate our reality with the planet has never been more articulated, and consequently starts our excursion of 'Clearing the Way Towards Economical Living.'

Picture a reality where each step you take, every decision you make, is a cognizant step towards a greener future. It's a reality where the air is cleaner, the seas are overflowing with life, and the timberlands reverberate with the tunes of animals extraordinary and little. The excursion we leave upon today is one of change and strength, an excursion that imagines a world that is not simply feasible

The Clarion Call of Supportability

The clarion call of supportability resonates through the halls of time, asking us to reconsider, rethink, and rebuild our lives. It's a challenge to leave the straight model of utilization and embrace a roundabout economy that reinvigorates the disposed of, reusing waste into important assets. This change in perspective is a demonstration of human resourcefulness and development. It's a guarantee to people in the future that the missteps of the past won't be rehashed, that we are creating a tradition of careful living.

The Three Mainstays of Manageability

At the core of our process lie the three mainstays of manageability: natural, social, and monetary. These support points structure a sensitive harmony, a trinity that should be sustained and maintained for an amicable world. Ecological supportability tends to our relationship with the normal world - the air we inhale, the water we drink, the scenes we navigate. It's a vow to safeguard biodiversity, decrease carbon impressions, and put resources into sustainable assets. Social maintainability dives into the multifaceted snare of human cooperations, supporting for inclusivity, civil rights, and equivalent open doors for all. Ultimately, financial supportability takes a stab at thriving without settling for less, underscoring capable strategic policies, fair exchange, and moral trade.

The Ensemble of Individual Decisions

Each decision we make, regardless of how immaterial it might appear, reverberates through the texture of maintainability. From the food on our plates to the garments in our wardrobes, every choice is a chance to decide in favor of a superior world. Picking plant-based dinners, embracing energy-proficient apparatuses, and supporting neighborhood craftsmans are strings in the embroidered artwork of feasible living. An ensemble of individual decisions all things considered blend into a resonating melody of progress.

Developments Molding Tomorrow

The excursion towards maintainable living is cleared with historic developments that reclassify enterprises and reshape standards. The quick headway of innovation has birthed another time of conceivable outcomes, from sun powered chargers that saddle the sun's energy to electric vehicles that murmur down the roads. Vertical cultivating is rethinking agribusiness in metropolitan scenes, diminishing food miles and supporting food security. Developments in material science are birthing biodegradable plastics, offering a promising option in contrast to their naturally hurtful partners. These developments are not only fantasy dreams; they are substantial arrangements that enlighten the street ahead.

Aggregate Activity: The Force of We

While individual decisions are the structure blocks of supportable living, aggregate activity is the cornerstone that maintains some kind of control. The force of networks meeting up with a common perspective is a relentless power that can reshape strategies, challenge enterprises, and reform outlooks. Grassroots developments upholding clean energy, zero-squander ways of life, and reforestation are switching things around against natural debasement. Legislatures and enterprises are likewise noticing the call, setting aggressive focuses to decrease discharges, monitor assets, and put resources into a greener future.

Education: Enlightening the Way Forward

Instruction remains as a reference point, directing us through the maze of supportable living. The key opens how we might interpret the complex connections between environments, the results of our activities, and the pathways to change. By outfitting people with information, we engage them to settle on informed choices that echo through society. Instructive establishments, local area studios, and computerized stages assume vital parts in dispersing data and supporting the heads of tomorrow.

The Difficulties We Should Survive

However, our excursion towards supportable living isn't without its difficulties. The idleness of laid out rehearsals, financial interests, and an absence of mindfulness frequently upset progress. Tending to these difficulties requires an adjustment of strategies, yet have a significant impact on in outlooks. About destroying the hindrances forestall the reception of economical works on, encouraging coordinated effort, and sustaining an aggregate liability regarding the planet.

The Commitment of a Greener Skyline

As we navigate the way towards economical living, we're not simply modifying the story of our lives; we're writing a commitment for the future.

A future where urban communities are lavish with green spaces, where seas are liberated from plastic tides, and where each residing being flourishes as one. The excursion is complicated, multi-layered, and frequently overwhelming, yet it's an excursion worth setting out upon. An excursion celebrates human potential, loves the miracles of nature, and champions the versatility of life itself.

Join the reason, embrace Eco-Care Fundamentals, and assemble the way toward maintainable living. Your choices count, and together, we can construct a greener, better future for our reality and ages to come. Act right now.

Chapter: 1

Understanding Sustainability: Principles and Concepts

1:1 Defining Sustainability: Beyond Environmentalism

In a world going through remarkable change, the expression "manageability" has developed from its hippie roots to turn into a complex idea enveloping the central center of human life. Manageability is not generally restricted to keeping up with biological systems or rationing assets; it progressively has implications in financial aspects, civil rights, culture, and, surprisingly, individual prosperity. We find an embroidery of entwined values that guide us towards a stronger and agreeable future as we dig into the powerful scene of characterizing manageability past environmentalism.

From Green Starting points to Comprehensive Viewpoints

Manageability was frequently connected with "practicing environmental awareness," the energizing cry of hippies trying to diminish contamination, cut carbon impressions, and preserve weak biological systems. While these are as yet significant parts of supportability, the thought has developed into a more extensive structure that perceives the perplexing association among nature and human culture.

At its establishment, manageability is the acknowledgment that each choice we have affects the world external ourselves. A source of inspiration expects us to go past our ongoing requirements and think about the drawn out results of our choices. Supportability urges us to proceed delicately on the world, perceiving that the planet's assets are limited and should be involved carefully to keep up with their accessibility for people in the future.

The Underpinnings of All encompassing Manageability

The three mainstays of maintainability are currently natural, social, and monetary. These support points act as the foundation for a tough society that makes progress toward an equilibrium of human prosperity, biological wellbeing, and monetary turn of events.

1. Natural Supportability: This point of support remains consistent with its biological starting points, featuring the meaning of safeguarding our planet's delicate environments. Ecological supportability stresses our obligations as Earth stewards, from keeping up with biodiversity and safeguarding regular living spaces to decreasing contamination and embracing environmentally friendly power sources.

2. Social maintainability: reaches out past natural worries to incorporate issues of civil rights and inclusivity. It requests equivalent asset assignment, the annulment of segregation, and the assurance of central basic freedoms to all people. Social manageability underlines the need of an only society in guaranteeing long haul strength and harmony.

3. Financial manageability: comprehends that financial frameworks should change to keep away from asset consumption and ecological mischief. Monetary manageability involves asset the executives that is capable, moral strategic approaches, and interest in firms that benefit society and the climate.

Strings of Supportability in Culture and Custom

Since supportability includes such countless various parts of life, it normally converges with culture and custom. Native information, for instance, is a wellspring of long haul customs that have supported environments for ages. Customary cultivating techniques, natural medication, and local area based asset the executives are completely founded on ideas of supportability. The undertaking is to consolidate these verifiable procedures with current headways to make a symphonious mix that answers present day concerns.

In the Supportability Condition, Individual Prosperity

A rising understanding in the drive for maintainability is that singular prosperity is inseparably attached to the prosperity of the world. A feasible way of life involves something beyond diminishing waste and preserving energy; it likewise involves developing mental, profound, and actual health. Care, honest utilization, and association with nature are turning out to be progressively significant parts of practical living. All things considered, a general public can't be genuinely supportable on the off chance that its individuals are worried, confined, or sick.

Instruction as a Change Impetus

To characterize maintainability in this more extensive point of view, training should go past natural information. It requires a familiarity with complex frameworks, as well as decisive reasoning skills and compassion. Instruction is the torchbearer, driving us from rushed ends and toward additional all encompassing arrangements. We give the structure to a general public that can adjust, improve, and flourish notwithstanding difficulties by developing an age of proficient residents.

Advancement and Innovation's Job

The computerized age has expanded mankind's capacity for development, opening up new ways for long haul progress. Shrewd urban communities advance energy use and work on metropolitan everyday environments by utilizing information and innovation. Environmentally friendly power, practical farming, and round economy ideas are altering areas. Development is overcoming any barrier between monetary development and ecological obligation, exhibiting that supportability isn't a hindrance to progress, but instead a way to it.

Organizations for Progress in Business and Manageability

Organizations, previously seen to be the reason for ecological harm, are presently viewed as problem solvers. The business area is embracing manageability as a direct idea, understanding that social and natural obligation isn't simply moral, yet additionally monetarily useful. Green practices, transparency in supply chains, and the fuse of feasible points into organization plans are changing enterprises and impacting client decisions.

The Supportable Development's Inclusivity

Past environmentalism, characterizing manageability involves understanding the relationship of worldwide worries. Environmental change, neediness, imbalance, and different worries are interwoven and affect various regions. To accomplish genuine maintainability, consideration should be focused on. To guarantee that nobody gets abandoned, arrangements should be adjusted to the singular prerequisites and real factors of various locales and socioeconomics.

Making an Enduring Heritage

Manageability is the string that associates the past, present, and future in the glorious embroidery of presence. A creating thought joins natural insight, social decency, and financial imperativeness. Past environmentalism, supportability is a way of thinking that urges us to envision a general public in which each individual's prosperity is sustained, societies flourish, and the earth flourishes. We become designers of a heritage that esteems the magnificence and intricacy of life in the entirety of its structures as we rethink supportability.

1.2 The Three Manageability Points of support: Natural, Social, and Financial Agreement

In an interconnected and convoluted world, the journey of maintainability has arisen as an encouraging sign and a way for an additional reasonable and impartial future. This objective is established on three support points: ecological, social, and monetary manageability. These three points of support involve the trinity that drives us to a quiet living with our earth and with each other. As we dig further into these support points, we reveal the confounded snare of connections that characterize the course to a more splendid, more economical future.

Ecological Supportability: Earth Watchmen

An extraordinary appreciation for our planet and its fragile biological systems sits at the core of the manageability development. Ecological maintainability is a devotion to moderating the regular habitat and the sensitive equilibrium it keeps up with. It recognizes that the World's assets are restricted and should be involved to keep up with their accessibility for people in the future.

Our biological systems are indispensable resources that help us, from the lavish rainforests rich with assortment to the immense seas that support life. Natural supportability involves measures for securing and reestablishing these environments, decreasing contamination, and lessening the effect of human action on the world.

Sustainable power sources, for example, sun based and wind power, embody ecological maintainability by bridging
the innate energy of the Earth without exhausting limited assets. Protection tries, ranger service projects, and

untamed life conservation endeavors all add to the honorable objective of safeguarding our regular legacy.

Strings of Inclusivity and Equity in Friendly Supportability

While ecological manageability is worried about the world,

social supportability is worried about individuals. It is the acknowledgment that a fair and impartial society is expected for an economical world. This point of support intends to advance consideration, uniformity, and civil rights for all individuals, no matter what their experience, character, or conditions.

Social manageability advocates for fundamental basic freedoms like clean water, good food, sanctuary, instruction, and medical care. It plans to kill separation, persecution, and precise incongruities that hold individuals back from achieving their maximum capacity. Social manageability makes the preparation for a quiet and strong world by fostering a general public in which each individual has the chance to flourish.

Local area commitment, strengthening, and variety advancement are basic parts of social manageability. Grassroots developments that supporter for underestimated populaces, orientation equity projects, and destitution battling regulation are ventures toward making a general public that qualities and safeguards the prosperity of every one of its individuals.

Financial Supportability: Finding Some kind of harmony Among Success and Obligation

Monetary supportability moves the financial scene away from benefit and toward the drawn out prosperity of the two individuals and the earth. It comprehends that monetary development should not come to the detriment of exhaustion of assets, ecological corruption, or social shamefulness.

Monetary manageability involves mindful asset of the executives, moral corporate practices, and speculations that focus on monetary returns as well as certain cultural effects. It is tied in with growing the meaning of achievement to incorporate monetary models, yet additionally friendly and ecological files.

toward a more purposeful methodology. It urges us to stop, contemplate, and gauge the compromises that accompany our choices. While certain choices might lean toward present solace, others might require the present penances in return for a more brilliant tomorrow.

Obligation and Inheritance

The idea of adjusting present and future requests is a string that sews together the texture of ages in the embroidery of presence. It fills in as an update that our choices have long haul suggestions that expand well past our nearby point of view. This rule urges us to procure consideration, intelligence, and a feeling of obligation that rises above time.

Allow us to recall, as we navigate the intricacy of current life, that our reality is characterized by the delights of the present as well as by the heritage we leave for people in the future. A reasonable methodology praises the present desires while recognizing the upcoming yearnings. By saving this sensitive equilibrium, we become stewards of a world in which abundance isn't just capable by us but on the other hand is woven into the texture of time itself.

Past Individual Impact: Decisions' Expanding influences, The idea of adjusting present and future requests stretches out past the singular level to incorporate the aggregate impact of our decisions. Think about metropolitan preparation, where choices about foundation, transportation, and asset conveyance have expansive ramifications. Urban areas that focus on productive public transportation and regular spaces work on the personal satisfaction presently as well as make the foundation for long haul metropolitan manageability.

Additionally, in the realm of innovation, development habitually shows up no sweat and speed. A drawn out vision, then again, requires thought of potential moral, natural, and cultural repercussions. Reception of new innovation without completely realizing its drawn out repercussions can bring about unforeseen unfortunate results. Dependable mechanical improvement requires gauging the possible impacts of our choices on both current and people in the future.

Despite Vulnerability, Flexibility

The idea of adjusting present and future requests advances flexibility, or the capacity to climate future vulnerabilities. A development that acknowledges a drawn out vision might beat unanticipated impediments with greater flexibility, much as a very fabricated structure expects changes in climate and climate.

Maintainable horticulture methodologies that focus on soil wellbeing and trim expansion show this flexibility. Ranchers secure a consistent food supply for quite a long time into the future via really focusing on the land today. Approaches that put resources into sustainable power sources and fiasco readiness, as a rule, fortify cultural versatility, reducing the effect of environment related disasters and financial shocks.

Manageable Utilization: A Drawn out Vision Support point

Utilization is a superb illustration of how adjusting present and future prerequisites might bring about sensational change. Maintainable utilization is a methodology that urges mindful decisions today to get future asset accessibility. It involves considering the natural effect of our buys and picking items that are durable, energy-productive, and morally made.

Economical utilization additionally involves diminishing waste using reusing, upcycling, and moderate standards. It raises doubt about the expendable mindset that portrays current industrialism, welcoming us to esteem higher standards without ever compromising and strength over out of date quality.

Administration and Strategy, Whether in government, business, or common society, administration is basic to fostering a drawn out objective. Leaders who underline people in the future's prosperity and remember reasonable practices for their technique set the vibe for positive change. They overcome any issues between current reality and future objectives by creating strategies that advance impartial development, ecological insurance, and social advancement.

Carbon valuing, outflow decrease goals, and motivators for environmentally friendly power reception are instances of arrangements that advance feasible practices and show the thought of adjusting present and future requests. These approaches perceive that transient financial increases should be adjusted against the planet's and its occupants' drawn out wellbeing.

Rising above Limits: Worldwide Citizenship, The idea of adjusting present and future requests crosses borders, inciting us to take on a worldwide citizenship point of view. In an interconnected world, choices made in one area can have sweeping outcomes. Environmental change, for instance, is a test that needs global collaboration in light of the fact that its effects rise above public limits. Worldwide environmental change endeavors, like the Unified Countries Manageable Improvement Objectives (SDGs), feature the need of a typical obligation to a practical future. We can intensify our effect and work cooperatively to create positive change that rises above a long ways past our nearby networks by joining forces across societies, countries, and thoughts.

Fostering a Liberality Outlook

The idea of adjusting present and future requests urges us to develop a liberal outlook. It propels us to contemplate the prosperity of others, especially people in the future, and to settle on choices that mirror that idea. We can possibly be stewards of the earth and the human local area, similarly as guardians put resources into their youngsters' schooling and prosperity for a more promising time to come.

Liberality reaches out past monetary gifts to our activities, choices, and aims. We leave a tradition of empathy and responsibility by pursuing choices that advance everyone's benefit. This heritage turns into a gift to people in the future, a recognition for our devotion to a world in which the quest for joy coincides with the commitment to safeguard what's in store.

Motivating Change Through Agreeable Equilibrium

The thought of adjusting present and future requests isn't an equation, yet rather a persistent excursion of reflection and care. It requires progressing consideration of the ramifications of our choices, as well as the readiness to make course changes when important. This powerful methodology perceives that the harmony between the present and what's to come is continually moving because of evolving conditions, new data, and developing qualities.

By embracing this thought, we become progressive change specialists. We embody the conviction that, when lined up with a drawn out vision, human demonstrations might modify the direction of society, economies, and biological systems. We honor the excellence of the present while making a heritage that reverberations through the corridors of time as we explore this confounded dance.

The tune of the present fits with the reverberations representing things to come in the orchestra of presence

Chapter 2

Environmental Stewardship: Preserving the Planet

2.1   Biodiversity   Conservation: Safeguarding Environments and Species,

Biodiversity sparkles as a brilliant pearl in the rich mosaic of life on The planet, addressing the huge assortment of species and environments that make our reality an embroidery of miracle. Biodiversity is something beyond a show; it is the fundamental foundation of human life. Each specie, from the littlest microscopic organisms to the biggest monster, plays an unmistakable part to act in the convoluted dance of biological systems. As we dig more into the captivating universe of biodiversity preservation, we understand that safeguarding these fragile strings of life that weave the texture of our world is so basic.

The Reliance and Equilibrium of Life, Think about biodiversity as a sensitive web, where each string addresses an animal category and every association mirrors a convoluted communication. Biological systems are the powerful outcome of these collaborations, where species cooperate with each other and with their environmental elements, bringing about a finely adjusted balance. Every species adds to the solidness and usefulness of biological systems, from pollinators that guarantee plant expansion to carnivores that keep up with populace control.

At the point when a solitary string is eliminated, it influences the whole web. This profound interconnectedness features the delicacy of biodiversity; the elimination of even apparently little animal categories can have flowing repercussions that change the equilibrium of an entire biological system. Biodiversity protection comprehends that the soundness of these interlinked frameworks is inseparably connected to our prosperity.

Biodiversity conveys an abundance of biological system administrations - benefits that nature presents to people, frequently without our insight. Clean air, clean water, rich soil, and environment, the executives are a couple of indispensable administrations for our reality and prosperity. Biodiversity safeguards these administrations by permitting biological systems to endure disturbances and adjust to changes.

Woodlands, for instance, work as carbon sinks, retaining and putting away CO_2, a critical ozone harming substance. Storm floods and disintegration are relieved by mangrove biological systems. The complicated trap of life in the seas manages the planet's environment and supplies a significant area of the planet oxygen. Biodiversity protection has become connected with saving these fundamental capabilities that act as the underpinning of our communitie

Variation and Advancement in Species Variety

The unbelievable variety of species on Earth exhibits the virtuoso of nature's transformative excursion. Every species is the climax of millions of long periods of variation and imagination, bringing about shapes and works that have dominated specialty step by step processes for surviving. Biodiversity is a living library of organic data, a mother lode of expected answers for the present and the upcoming hardships.

Consider a few animals' remarkable capacity to prosper in cruel settings. The variety of species presents the chance of imaginative experiences for disciplines like wellbeing, farming, and innovation, from extremophiles living in the cruelest conditions to plants making particular substance compounds for guard. Biodiversity assurance guarantees that we keep on approaching this fortune of information to help humankind.

While biodiversity might be found in each side of the globe, a few regions stand apart as biodiversity areas of interest - regions that have a very big number of species, a considerable lot of which are found no place else. These areas of interest are essential organic variety repositories, however they are additionally excessively delicate to human exercises like natural surroundings corruption, contamination, and environmental change.

The Amazon rainforest, for instance, is popular for its monstrous biodiversity, which incorporates incalculable plant, creature, and organism species. In any case, unreasonable deforestation takes steps to destroy this awesome embroidered artwork of life. Biodiversity security becomes basic in these areas of interest in light of the fact that the elimination of species here can have broad worldwide results.

The 6th Mass Eradication: An Admonition, The Earth is presently encountering the 6th mass termination - a quick loss of animal categories like the dinosaur eradication episodes. Not at all like prior annihilations, this one is being driven basically by human movement. Contamination, over-abuse, and environmental change are adding to an unrivaled loss of biodiversity.

Preservation of biodiversity is critical to rectifying this pattern. We can turn around the downfall of species and environments by keeping up with and reestablishing territories, carrying out manageable ways of behaving, and bringing issues to light. The 6th mass eradication goes about as a reminder, requesting that we reevaluate our association with nature and find serious ways to save it.

The Worth of Nature: Morals and Feel, Biodiversity security reaches out past environmental need to incorporate moral contemplations and social importance.

Native people groups, for instance, every now and again have solid otherworldly and social connections to explicit creatures and natural surroundings. Biodiversity likewise has stylish worth, inspiring craftsmen, rationalists, and intelligent people over the entire course of time to concentrate on the regular world's magnificence and intricacy.

Preserving biodiversity involves something beyond keeping up with individual species; it additionally involves recognizing the natural significance of all types of life. About commending the variety has advanced over centuries, the ensemble of presence that fills our globe with amicability, variety, and astonishment.

The Job of Insurance Endeavors: Confidence in real life, Luckily, there is hopefulness in the field of biodiversity security. Protection exercises drove by individuals, gatherings, and legislatures are having an effect. Public parks and untamed life saves, for instance, give places of refuge to species to flourish. Protectionists work constantly to restore debased environments, once again introduce imperiled species, and anteroom for strategy changes that esteem biodiversity.

Worldwide drives, like the Show on Organic Variety, unite nations to set biodiversity preservation and maintainable advancement objectives. These drives comprehend that our current circumstance is interconnected and that the strength of species and biological systems is inseparably connected to the wellbeing of humanity.

Biodiversity security isn't the occupation of a limited handful; an aggregate liability falls on us all. Each individual can settle on choices that assistance to safeguard biodiversity. Our exercises affect the trap of life, from elevating supportable ways of behaving to battling for more grounded protection guidelines.

By tolerating the idea of biodiversity assurance, we become stewards of the normal world, species guardians, and biological system safeguards. We outline a course for a future in which the wonderful texture of life keeps on astonishing, dumbfound, and support us. Allow us to recall that by keeping up with the variety of life, we are likewise safeguarding our inheritance and guaranteeing a world that is rich, strong, and lovely for people in the future.

Environment Administrations: Nature's Gift to Humankind, The idea of biodiversity security is extended through the significant administrations given by environments. These environmental administrations are in some cases underestimated since they capability behind the scenes, quietly supporting our reality in manners we may not completely appreciate. Environments work as inconspicuous gatekeepers of our prosperity, from the cleaning of air and water to the guideline of environment and disease.

Wetlands, for instance, work as regular water channels, sifting through poisons and impurities from water sources. Timberlands play a significant capability in retaining CO_2 and delivering oxygen, which assists with decreasing environmental change. Coral reefs shield beach front towns by engrossing the power of waves and tempests, safeguarding them from the damaging impacts of catastrophic events.

Preservation of biodiversity guarantees the continuation of these basic natural capabilities. We protect the multifaceted snare of connections that supports the Earth and its occupants by shielding differed species and their environments.

The Moral Goal: Safeguarding All Living things, Biodiversity protection is unequivocally established in moral worries that perceive all life structures' natural worth. This perspective goes against the human-centric view that places mankind at the focal point of the normal world, guaranteeing that all species reserve an option to reside no matter what their utility to people. Biodiversity protection, similar to basic freedoms and civil rights, advocates for the privileges of nonhuman creatures.

The elimination of an animal categories is a social and moral misfortune as well as a logical misfortune. Every species addresses a particular section in the immense story of life, adding to our planet's magnificence and variety. Biodiversity protection welcomes us to recognize the characteristic worth of all types of life and our obligations as stewards of a different and interconnected climate.

Native Insight and Conventional Environmental Information

For quite a long time, native gatherings all over the planet find lived as one with their surroundings, laying out a profound consciousness of the environments they occupy. Conventional natural information, which has been gone down through oral customs, gives bits of knowledge into supportable practices, asset the executives, and the intricate cooperations between species.

Biodiversity protection is tied in with gaining from the insight of people who have coincided with nature for centuries, not simply logical ability. Native methods like rotational cultivating, land stewardship, and comprehensive asset the executives show the agreeable living together that people and the climate are prepared to do. We fortify our endeavors to shield biodiversity and secure a reasonable future by regarding and consolidating native environmental information.

Monetary Worth: The Secret Advantages of Biodiversity, Biodiversity protection is something beyond a natural concern; it additionally has significant financial implications. Biological systems support economies by providing assets like food, wood, and medication. Also, the eco-the travel industry and sporting exercises like bird-watching and untamed life safaris produce huge cash and open positions.

The deficiency of biodiversity subverts these monetary advantages. Overfishing, for instance, can make fisheries breakdown, influencing the jobs of millions. Biological system corruption can likewise prompt diminished horticultural efficiency and expanded weakness to catastrophic events. Perceiving the financial meaning of biodiversity reinforces the case for its conservation and advances a drawn out balance between natural wellbeing and monetary turn of events.

The Job of Innovation: Protection Development, In the computerized period, innovation is a gigantic partner in the battle to save biodiversity. Satellite photography and remote detecting, for instance, empower us to follow changes in biological systems and feature areas of concern. In any event, when actual perception is troublesome, DNA sequencing can assist with distinguishing species.

Prescient demonstrating, empowered by creative innovation, for example, man-made reasoning and AI, can illuminate preservation strategies. Drones are utilized to screen natural life populaces and gather information in hard to-arrive districts. We can reinforce our endeavors to keep up with biodiversity and pursue taught choices that safeguard the regular habitat by utilizing the force of innovation.

Instruction and Promotion: Engaging Change-Creators, instruction, and support are basic drivers of biodiversity preservation. Raising information on the worth of biodiversity, its dangers, and people's parts in its assurance is basic for coordinating aggregate activity. Schools, foundations, and local area associations can all have a significant impact in imparting a protection culture in kids since the beginning.

People can become biodiversity preservation advocates by giving to protection associations, taking part in local area ventures, and standing up to request strategy changes. Web-based entertainment stages are a compelling method for bringing issues to light and rousing activity on a worldwide scale. By giving information and assets to change-producers, we make a development that waves through networks, enterprises, and states.

A Source of inspiration: Biodiversity Protection, Biodiversity preservation is certainly not a theoretical idea; a source of inspiration requires our consideration, aim, and devotion. Each person, whether by private decisions, neighborhood drives, or worldwide collaboration, plays a part to play in this endeavor. Our Earth is a living ensemble, with every species contributing an unmistakable note to the tune. At the point when one note is dropped, the amiability is perplexed.

Allow us to recall that we are essential for this orchestra, with the capacity to make a more promising time to come. Protection of biodiversity is a gift to ourselves and people in the future, a tradition of stewardship and obligation. Allow us to hold in our souls the excellence of a planet overflowing with life as we navigate the difficulties of the cutting edge world, and let our activities mirror a significant devotion to defend and keep up with the mind boggling snare of presence that supports every one of us.

2.2 Asset The executives: Utilization and Waste Decrease

The craft of asset the board arises as a basic light of manageability in a general public portrayed by restricted assets and rising individuals. A direct yet significant idea urges us to capitalize on what we have while restricting waste and lavishness. We find the force of smart choices, the importance of preservation, and the changing effect it holds for the earth and people in the future as we dive into the area of asset the executives.

The Dance of Equilibrium: Mindfully Addressing Needs, Asset the executives is a fragile dance of equilibrium. It is the acknowledgment that the World's abundance is limited and that each asset we use has a critical effect. This thought urges us to move toward utilization carefully, considering our quick necessities as well as the planet's drawn o ut prosperity.

It's anything but a question of lack or hardship; rather, it involves overseeing assets morally and effectively. From energy to water, natural substances to food, asset the executives expects us to settle on choices that safeguard the delicate equilibrium of our environments. It's a harmonious connection between our longings and the planet's capacity to give.

In a customer driven world, asset executives compel us to reexamine our relationship with material things. It's a way to focus on better standards without compromise, to put resources into things that last instead of sustaining a dispensable culture. Careful utilization involves valuing things that are durable, repairable, and don't add to piles of waste.

Consider the effect of quick design, which is the fast production of minimal expense clothing that much of the time winds up in landfills after a couple of years. Clothing made from reasonable materials, delivered under moral circumstances, and intended for life expectancy is energized by asset the board. It addresses a change from an expendable culture to one of purposeful stewardship.

Energy productivity: From preservation to renewables, energy utilization is essential to asset the board. While petroleum products were once the foundation of our energy frameworks, they have harmed the climate. Energy preservation is empowered by asset the board strategies, for example, switching out lights when not being used, further developing warming and cooling frameworks, and using energy-productive gear.

A definitive change, be that as it may, rests in the shift to environmentally friendly power sources. Sunlight based, wind, hydro, and geothermal energy use nature's power without draining limited assets or creating perilous poisons. Perceiving the capability of these other options and upholding for their broad reception is what's truly going on with asset the board.

The Food Association: Squander Decrease, Food is more than just a wellspring of nourishment; likewise a significant asset requires legitimate administration. 33% of all food created on the planet is squandered, adding to hunger as well as natural destruction. Food asset the board involves diminishing waste through procedures, for example, feast arranging, fertilizing the soil, and supporting food recuperation programs.

It's likewise about choosing food sources with a more modest ecological effect. Taking on a plant-based diet disposes of the requirement for asset concentrated creature cultivation. Asset the board is like faithful eating in that it includes settling on choices that sustain our bodies as well as the world.

Decrease, reuse, and reuse are the three R's.

"Diminish, reuse, reuse" typifies the embodiment of asset the executives. These three stages comprise a guide to dependable utilization and waste decrease.

The initial step is to diminish our admission. It involves gauging our requirements against our longings, wiping out pointless pressing, and choosing items with low natural effect. Utilization decrease reaches out to water use, transportation, and different parts of day to day existence.

Reuse: Rehearsing reused items assists with expanding their life expectancy. Reusing items moderates assets and limits the interest for new assembling, from utilizing fabric sacks rather than plastic ones to reusing holders. It is a social change from an expendable mentality to one of innovativeness and development.

Recycle: The third mainstay of supportable assets the board is reusing. It involves redirecting trash from landfills and returning materials to the assembling system. Viable reusing, then again, requires sufficient material partition, familiarity with neighborhood reusing projects, and backing for exercises that make a roundabout economy.

Asset the board reaches out past individual exercises to embrace whole financial models in the progress from direct to round economies: Changing Ideal models. The normal direct economy of take, make, and arrange exhausts assets and produces a lot of waste. A round economy, then again, expects to diminish squander and keep up with wares available for use.

Items in a roundabout economy are solid and fixed. Materials are reused and reused, bringing down the interest for unrefined components. Reusing is focused on, and squander is decreased through shrewd plan and upright utilization. The roundabout economy is embraced by the executives as a guide for long haul improvement.

Training turns into a foundation of asset management because of **the instructive objective:** Sustaining Outlooks. Enabling individuals with data about ecological impacts, the significance of feasible exercises, and the repercussions of waste gathering encourages a capable mentality. Schools, people groups, and associations may all have a significant impact in bringing up ecologically cognizant youngsters.

People can be instructed about the significance of asset protection, the results of their decisions, and useful waste-decrease systems through instructive projects. We lay out a tradition of learned leaders who are ready to confront the troubles of asset executives by conferring these standards early in life.

The Job of Innovation: Asset The executives Development, In the advanced age, innovation opens up new vistas for asset the board. Savvy gadgets, for instance, consider the checking and improvement of energy utilization. Applications and stages unite networks to share assets and diminish squander. Sensors screen squander levels in canisters, permitting waste assortment courses to be enhanced and outflows diminished.

Moreover, innovation works with the worldwide trade of thoughts and arrangements. Online stages support feasible ways of behaving, examples of overcoming adversity, and gathering activity. Innovation is a driver of development and positive change in the field of asset management.

From Nearby to Worldwide Effect: Aggregate Liability, Asset The executives Cross Geological Limits. A standard unites people, networks, companies, and states to cooperate to safeguard the planet's assets. While individual demonstrations are significant, enormous scope impact requires coordination and strategy changes at the government and state levels.

Global deals, like the Unified Countries Economical Improvement Objectives, stress the meaning of capable asset executives. These arrangements give a way to countries to cooperate toward a more populist and environmentally adjusted future by characterizing focuses for feasible utilization and creation.

Asset the executives is definitely not a troublesome limitation in making a feasible heritage; an enabling disposition welcomes us to be stewards of the World's wealth. It's undeniably true that our choices shape the world; the assets we monitor currently lay the foundation for a more maintainable future. We become planners of a heritage characterized by equilibrium, regard, and flourishing for all by taking on insightful utilization, limiting waste, and pushing for dependable practices. As we walk cautiously on this planet, let us recollect that asset the board is an honor and a potential chance to make a world that flourishes for people in the future.

Feasible Advancement: Reclassifying Progress, Asset The board is a valuable chance to rethink progress, not a call to balance. Monetary development has generally been surveyed simply regarding GDP (Gross domestic product), frequently overlooking natural and social ramifications. The idea of a reasonable turn of events - a complete methodology that considers financial, natural, and social prosperity - goes against this limited viewpoint. Inside this worldview, maintainable development arises as a directing light. It is the mindfulness that improvement might be made while regarding the planet's limits through inventiveness, innovation, and business. Green advances, roundabout plans of action, and practical metropolitan arranging are instances of human resourcefulness working together as one with nature.

Local area Commitment: Grassroots Change, When people group are involved and engaged, asset the board flourishes. Neighborhood drives driven by residents, organizations, and associations have huge waste decrease and preservation potential. Local area gardens, treating the soil projects, and zero-squander developments all originate from a common craving to impact positive change.

Local area commitment advances reasonable responses as well as a feeling of having a place and shared liability. It fills in as an update that asset the executives is an excursion that lives on the strength of collaboration, shared information, and a shared objective for a superior future.

Forming Enterprises Through Corporate Obligation

Enterprises are basic to asset executives. Rethinking strategic approaches to line up with practical qualities is essential for corporate obligation. Taking on eco-accommodating assembling procedures, diminishing energy use, and advancing the utilization of inexhaustible assets are all important for the arrangement. It is additionally critical to embrace moral inventory chains that regard common liberties and the climate.

Organizations that focus on asset the executives not just assistance to make the world a better spot yet additionally work on their standing and appeal to ecologically concerned clients. Reasonable practices become a brand name of ground breaking firms, laying the foundation for a rewarding and manageable business landsca

Strategy and Guideline as Change Specialists

Strategy systems and regulations are expected to advance successful asset the executives. States assume a significant part in cultivating a climate that supports manageable practices while putting inefficient direct down. Strategies could go from burdening single-use plastics to requiring reusing programs and laying out energy effectiveness guidelines.

Global deals, like the Paris Understanding and the Maintainable Improvement Objectives of the Unified Countries, lay out an overall plan for asset management and natural security. These deals lay out a system for states to team up toward a more feasible future, perceiving that the test of asset the board rises above limits.

Squander Decrease: A Way to Zero Waste

Squander decrease and asset the board are inseparably connected. The way to zero waste is decreasing how much waste we create and upgrading asset utilization. It involves choosing items with little bundling, supporting firms that esteem manageable practices, and lobbying for expanded maker obligation.

Embracing the idea of "upcycling" - reusing disposed of materials into new and important merchandise - is likewise essential for squander decrease. Upcycling keeps trash out of landfills as well as cultivates imagination and creativity in tracking down new applications for existing assets. Upcycling delineates the force of assets of the executives to advance great change, from reused furniture to design fabricated from reused materials.

Instructive Missions: A Lifestyle choice Intentionally

The effect of asset executives stretches out past the substantial; it pervades into our cognizance, impacting how we experience the climate. Crusades that underscore the meaning of dependable utilization and waste decrease prepare for cognizant living. They question standards, begin discussions, and rouse individuals to rethink their decisions.

Instructive endeavors can take many structures, going from motion pictures that draw light on the repercussions of overconsumption to school programs that show understudies harmless to the ecosystem ways of behaving. These projects urge people to go with choices that add to a more supportable future by bringing issues to light and obligation.

Flexibility Even with Misfortune

The board is laden with challenges. Quick urbanization, overpopulation, and environmental change all present troublesome difficulties. Notwithstanding, despite difficulty, the board fills in as a guide of flexibility. It persuades us to think innovatively, adjust to evolving circumstances, and construct a creative demeanor.

Water shortage, for instance, is a significant worldwide concern. Water protection, the improvement of successful water system frameworks, and the usage of elective water sources are undeniably advanced by asset executives. Asset the executives helps social orders in turning out to be stronger and versatile by standing up to troubles head-on and proactively.

Future-Arranged Strengthening

Asset the board traverses ages, as the choices we make today have long haul results. By taking on this methodology, we become designers of a future in which assets are esteemed, squander is diminished, and abundance doesn't come to the detriment of the planet. It's a tradition of strengthening, a gift to people in the future, an assurance that they, as well, will acquire a world brimming with potential outcomes.

Asset the executives is a way of cognizance and obligation, not an objective. It is a vow to offset our needs with the limit of the world. As we cross the complexities of current life, remember that asset the executives is an idea that offsets human advancement with the prosperity of the Earth. By taking on this mentality, we develop a future where riches and supportability coincide, and where each choice prompts a world that twists.

2.3 Environmentally friendly power: Tackling the Force of Nature

In a world adapting to the rising worries of environmental change and diminishing petroleum product assets, the consideration has immovably centered around sustainable power sources as the solution to our energy misfortunes.

Sustainable power, otherwise called "nature's power," has emerged as a wellspring of hopefulness, promising a cleaner, more maintainable future. This paper goes into the exciting universe of environmentally friendly power, looking at its different structures, amazing advantages, and imperative job in changing our planet for people in the future.

Sustainable power's Ascent

The historical backdrop of sustainable power might be followed back to old civilizations that pre-owned breeze and water power for crushing grain and water system. In the cutting edge period, we are nearly an energy upheaval, filled by another comprehension of science and innovation. Sun based, wind, hydropower, geothermal, and biomass energy have arisen as the foundations of this upheaval, giving a scope of economical options in contrast to petroleum derivatives.

Sun oriented Energy: Sparkling in the Sun

Sunlight based energy is one of nature's most enthralling appearances of force. The sun, an endless combination reactor, radiates a surprising measure of energy that far surpasses mankind's energy necessities by significant degrees. Sun powered chargers can now change over daylight into power all the more productively on account of advances in photovoltaic innovation. Sun based energy's compass is extending, from tremendous sun powered ranches in deserts to housetop establishments in metropolitan conditions, giving energy autonomy and bringing down carbon impressions.

Wind Energy: Riding the World's Breath

Another great exertion is outfitting the dynamic energy of the breeze. Wind turbines, which look like advanced windmills, convert the power of the breeze into power. The excellence of these designs rests in their capacity to gather energy from a passing regular power. Seaward wind ranches, in which turbines outfit the force of the vast ocean, have emerged as an original method for producing practical energy while diminishing area use. These transcending sentinels of maintainability are turning out to be more productive and cheap as innovation creates.

Hydroelectric Energy: Power Streams Diverted

For centuries, streaming water has been saddled, however current hydroelectric power offices take this idea higher than ever. Potential energy is held and afterward delivered to drive turbines, producing power, by damming streams and building repositories. The delightful Hoover Dam in the US and the Three Crevasses Dam in China are demonstrations of hydroelectric power's gigantic impact. This environmentally friendly power source gives a predictable power supply as well as assists with flood control and water the board.

Geothermal Energy: Tackling the World's Inward Intensity

A huge repository of intensity exists underneath the World's surface, ready to be gotten to. Geothermal energy produces power by tackling the innate intensity of the planet's inside. Profound wells are exhausted to get to high temp water and steam repositories, which are consequently used to turn turbines and produce energy. Iceland, arranged on a geographically dynamic district, has embraced geothermal energy with enthusiasm, showing its capacity to supply supportable power and in any event, warming answers for homes and organizations.

Biomass Energy: Nature's Reused Power

Drawing from natural assets like wood, rural squanders, and trash, biomass energy reflects the normal carbon cycle. We close the circle on fossil fuel byproducts by consuming these materials or changing over them into biofuels, making this energy source almost carbon-nonpartisan. Moreover, biogas produced from natural waste in landfills or anaerobic digesters assists with diminishing ozone harming substance outflows while additionally giving an energy source.

Benefits Other Than Power Age

The charm of sustainable power extends a long ways past the domains of force age. Its benefits cut across monetary, natural, and social aspects, introducing a change in perspective by they way we ponder supportable turn of events.

Ecological Stewardship: Sustainable power has unmatched ecological advantages. Renewables, in contrast to petroleum derivatives, produce practically zero nursery gasses, diminishing the emanations that cause environmental change. This reduction in air contamination brings about better air quality, lower medical care costs, and the conservation of biological systems and biodiversity.

Energy Security: Depending on limited petroleum derivatives opens nations to international contentions and supply interferences. Utilizing sustainable power sources expands the energy blend, further develops energy security, and diminishes international disagreements regarding assets.

Work Creation and Financial Development: The sustainable power business is growing quickly. The business creates a plenty of valuable open doors, empowering neighborhood financial development, from innovative work to fabricate, establishment, and support.

Development in Materials Science, Energy Stockpiling, and Network The executives: The mission for sustainable power has encouraged state of the art progresses in materials science, energy capacity, and framework the board. These advancements further develop sustainable power advances as well as have applications in different ventures, impelling by and large specialized progress.

Power Decentralization: Sustainable power permits individuals, networks, and countries to assume command over their energy creation. Roof sunlight powered chargers, limited scope wind turbines, and off-matrix gadgets give independence in energy, limiting dependence on unified energy sources.

Worldwide Energy Access: In remote or underserved regions with restricted admittance to regular power sources, environmentally friendly power offers an opportunity to jumpstart the customary energy framework, giving power to the people who need it most.

The Issues and the Way Forward

While environmentally friendly power holds incredible commitment, it isn't without its snags. Mix into current energy frameworks, discontinuity worries with sun powered and wind, and the prerequisite for compelling energy stockpiling choices are obstructions to boundless reception. Be that as it may, diligent innovative work are quickly eliminating these road obstructions.

Energy Capacity: Productive and practical energy stockpiling innovations are expected to keep a consistent inventory of force from irregular sources, for example, sunlight based and wind. Battery innovation progressions, siphoned hydro capacity, and arising advancements, for example, hydrogen capacity are pushing the environmentally friendly power transformation forward.

framework modernization: The ongoing energy framework was worked for concentrated petroleum derivative based power age. The matrix should be refreshed to take into account bidirectional energy move and brilliant administration of decentralized inexhaustible sources.

Strategy and Venture: State run administrations and financial backers are basic drivers of sustainable power organization. Steady arrangements, sponsorships, and consumptions in Research and development are significant to rushing the change away from petroleum derivatives.

Public Mindfulness and Schooling: It is basic to teach people in general about the advantages of sustainable power and expose false impressions to make far and wide acknowledgment and cooperation in the change.

Environmentally friendly power's ascent shows mankind's capacity to create and adjust. We are releasing nature's gigantic energy potential, from tackling daylight to catching the force of the breeze, streams, and Earth's center. This progress is something beyond a change in energy sources; it is a groundbreaking excursion toward a more practical, strong, and amicable future.

Environmentally friendly power is an image of confidence in a time of natural worries, promising a general public fueled by spotless, limitless sources. As innovation advances and mindfulness develops, the energy behind sustainable power develops, giving a more promising time to come to people in the future. It depends on us to pursue this open door and construct our existence by embracing the boundless energy that nature so liberally offers to us. The decision is clear: a future controlled by the sun, wind, water, and Earth, where nature's power impels us toward a greener skyline.

2.4 Environment Activity: An Invitation to battle Notwithstanding An unnatural weather change

In the expansive texture of mankind's set of experiences, there comes when our aggregate decisions have extensive outcomes a long ways past our lifetimes — when we stand at a junction with the capacity to change the destiny of our planet. This is an ideal opportunity to resolve the gigantic issue of an Earth-wide temperature boost with unfazed conviction and a firm source of inspiration. Environmental change is presently not an approaching risk; a reality requires our quick consideration, expecting us to limit its belongings through an ensemble of extremist measure

First Gander at the Environment Emergency

Believe the Earth to be a fragile environment in which every component coincides as one. Think about this harmony under attack: an orchestra hindered by increasing temperatures, modifying weather conditions, and vanishing ice sheets. This is the environment emergency, which is brought about by human exercises that discharge nursery gasses into the air, catching intensity and causing a chain response of unfortunate fiascoes. The need to act isn't just about safeguarding the earth; it's likewise about guaranteeing a tenable world for people in the future.

The Study of An Earth-wide temperature boost, Glancing through the logical focal point uncovers the different elements of an unnatural weather change. Nursery gasses like carbon dioxide and methane act as a sweeping around the Earth, catching intensity and raising temperatures. This is certainly not a characteristic event; it is the consequence of human activities like the utilization of petroleum derivatives, deforestation, and modern tasks, which discharge these gasses in exceptional amounts. The most vital move toward creating effective cures is to grasp the science.

The Arrangement Ensemble

An unnatural weather change requires an ensemble of procedures traversing areas, countries, and ages. Each note, from strategy changes in accordance with individual activities, adds to the congruity of a feasible future.

A Perfect Energy Blast

The rhythm of clean energy — a crescendo that integrates inexhaustible sources, for example, sun powered, wind, and hydropower — beats at the center of environment activity. This progress away from petroleum products decreases emanations while at the same time introducing another period of energy freedom, development, and occupation development. The perfect energy crescendo is in excess of a substitute; it's a potential

Moving to another lane in Feasible Transportation

Envision roads loaded up with delicately skimming electric cars, abandoning cleaner air and calmer networks. Feasible versatility is something beyond a change in gears; it is an excursion toward electric vehicles, proficient public transportation, and person on foot cordial metropolitan scenes. We can guide the boat toward a greener future by cutting emanations from transportation, the single biggest supporter of ozone depleting substance discharges.

Reforestation: Taking Life Back to the Earth

Believe trees to be the World's lungs, engrossing CO2 and ousting oxygen. Reforestation is a helpful interaction that sequesters carbon as well as jelly biodiversity and safeguards basic biological systems. With each tree planted, we revamp our planet's story, constructing a story of versatility and recovery.

The Roundabout Economy: Making Life span

Acknowledge the idea of a roundabout economy — a cooperative framework wherein assets are consumed, reused, and reused in a perpetual circle. The roundabout economy diminishes the weight on the World's assets by decreasing waste and limiting natural substance extraction while encouraging advancement in reasonable creation and utilization.

Aggregate Activity: One Voice

The battle against an Earth-wide temperature boost is certainly not a solitary one; it's a collective endeavor. Aggregate activity expands our impact and sends a strong message: the weep for change is too clear to even think about, from environmental strikes driven by enthusiastic teens to peaceful accords that interface countries together.

Transformation and Versatility: Exploring the New Ordinary

As we face the undeniable outcomes of environmental change, flexibility and transformation become our core values. Building versatile networks equipped for enduring outrageous climate occasions and adjusting to new truths is a basic part of environment activity. It is a promise to safeguard weak populaces, reinforce foundation, and plan for future issues.

Uncovering Examples of overcoming adversity from Advancements In the works

Environment activity seedlings are growing the whole way across the world. Creative advancements are changing over garbage into energy, afforestation projects are recovering parched places, and sustainable power establishments are arriving at new levels. These victories go about as encouraging signs, enlightening the changing force of our joint endeavors.

Altering Attitudes: A Change in outlook

Past actual strategy and mechanical forward leaps, environment activity requires a psychological shift — an enlivening to the interconnection of all life on The planet. It's an affirmation that our decisions, whether in utilization or protection, have broad results. This change in context is something other than a source of inspiration; it addresses a key change in our association with the earth.

The Gradually expanding influence: From Neighborhood to Worldwide, Consider a stone thrown into a lake, which makes swells spread all over. Environment activity is comparative in that a neighborhood drive can have a worldwide impact. A huge amount of energy, from diminishing individual plastic use to pushing for environmentally friendly power strategies, has an expanding influence that urges others and adds to the more noteworthy development toward manageability.

Cross-Boundary Advancement: Worldwide Coordinated effort

Environmental change has no limits, and its responses should be boundless. Worldwide collaboration is basic in battling a dangerous atmospheric deviation. Arrangements, for example, the Paris Understanding join countries by making promises to cut outflows and breaking point temperature climb. This coalition shows humankind's capacity to cooperate for everyone's benefit.

The Call of Youth: Gatekeepers Representing things to come

Think about a melody of youthful voices, enthusiastic, roused, and enduring in their call for environmental activity. Youth activists are at the very front of progress, talking truth to drive and catalyzing activity. Their energy advises us that the fight against a worldwide temperature alteration is about more than basically the present — it's about the inheritance we leave for people in the future.

Nature as a Companion: Biological system Reclamation

Nature is a powerful partner in environmental activity. From mangrove recovery to wetland restoration, environment rebuilding is a significant procedure for sequestering carbon, moderating environment influences, and safeguarding biodiversity. These regular cures balance a dangerous atmospheric deviation as well as reinforce our planet's capacity to endure its ramifications.

Local area Strengthening: Grassroots Developments

The battle against an Earth-wide temperature boost is complex, with networks at its heart supporting change starting from the earliest stage. Grassroots developments empower people to deal with their nearby surroundings by pushing for reasonable ways of behaving, bringing issues to light, and encouraging a cognizant culture that spreads

Industry Reshaping: Business Unbound

Enterprises of many kinds are taking on maintainable works on, perceiving that benefits don't need to come to the detriment of the climate. Ground breaking associations are redoing supply chains, embracing round economy models, and integrating sustainable power to clear another course toward capable development, from design to food creation.

Activism Through Workmanship: An Imaginative Unrest

Workmanship can get through boundaries and flash cultural change. Environment activism is no special case, with craftsmen utilizing their imaginative articulation to bring issues to light of natural difficulties, touch off discoursed, and evoke feelings that propel individuals to act. The imaginative upheaval, from visual craftsmanship to music, underlines the meaning of narrating in catalyzing change.

Moral Industrialism: Buying Power

Each purchase we make addresses a decision in favor of the kind of world we need to live in. Moral industrialism bridles the force of decision, empowering individuals to help items and organizations that stick to supportable strategies. Purchasers can affect market patterns and lift interest for harmless to the ecosystem options by settling on intentional choices.

The Heritage We Make: Pushing Ahead, the tradition of environment activity is something beyond decreasing emanations; it is additionally about the story we build for what's to come. It is tied in with leaving an existence where environments flourish, networks flourish, and people in the future acquire a world that is strong, lively, and in offset with nature. Our excursion to battle an Earth-wide temperature boost isn't simply a section; it's the underpinning of a centuries-boring tale.

As we finish up our investigation of environment activity, we perceive that it is an ensemble — a complex structure of endeavors, arrangements, and desires that cooperate to make a future we can be glad for. Each part of this ensemble, from the direness of the emergency to the flexibility of networks, is a challenge to take part, speak loudly, and have our impact in the most terrific of every single human undertaking: to guarantee the endurance of our planet and the flourishing of every one of its occupants. The stage is set, and we currently hold the mallet.

Chapter 3

Social Equity :(Empowering Communities Through Social Equity)

3.1 Inclusivity and Variety: Equivalent Open door

Variety is the string that fastens together the clear mosaic of societies, encounters, and viewpoints that enhance our reality in the embroidery of human life. No two individuals are precisely indistinguishable, similarly as no two snowflakes are indistinguishable. This natural variety is a wellspring of solidarity, development, and progression, and when fed by means of inclusivity, it makes the way for evenhanded chance for all.

The Impact of Variety

Consider a world where each blossom in a nursery is a similar tone, each bird sings a similar tune, and each tree bears similar organic products. An exhausting presence would fail to measure up to our normal world's dynamic and sensational excellence. Likewise, in human culture, variety fills in as a boost for imagination and development.

Different groups and networks bring assorted thoughts, encounters, and capacities to the table. This blend of differentiations invigorates development and the creation of groundbreaking thoughts. At the point when individuals from various foundations cooperate, they carry new viewpoints to critical thinking and independent direction. As indicated by a McKinsey and Company study, associations with different chief sheets are 21% bound to accomplish better than expected productivity.

Besides, assortment advances sympathy and expands skylines. Openness to assorted societies and perspectives challenges biases and advances liberality. It improves instruction by permitting understudies to draw in with a different scope of perspectives and advances a more profound information on worldwide interconnection. Inclusivity amplifies these benefits by guaranteeing that all voices are heard, giving a discussion to the underrepresented to communicate their accounts and add to aggregate progression.

Inclusivity as a Pathway to Rise to A potential open door, While variety supplies the natural substances, consideration is the recipe that changes those unrefined components into a flavorful dish. Inclusivity is something other than being addressed; it is the conscious act of laying out a climate where each part feels appreciated, esteemed, and enabled to contribute their best. About separating the limits hold individuals back from arriving at their maximum capacity.

Comprehensive associations and networks figure out that every individual adds something particularly amazing to the table. They forcefully search out and celebrate contrasts in orientation, race, nationality, age, sexual direction, religion, or capacity. This affirmation is a basic move toward killing underlying inclinations that have generally mistreated specific networks.

Representatives in a comprehensive work environment, for instance, have a solid sense of security offering their viewpoints unafraid of counter. This mental solace advances cooperation and innovativeness. It likewise cultivates a feeling of having a place, which is essential for self-improvement and satisfaction. As per Deloitte research, comprehensive groups beat their adversaries in group based evaluations by up to 80%.

Inclusivity is likewise an inspiring component behind equivalent instructive open doors. At the point when instructive organizations focus on cultivating a climate in which each understudy feels acknowledged and empowered, commitment increases and dropout rates fall. This, thus, sets out open doors for youngsters who may somehow get lost in the noise. It isn't only a question of selecting a different understudy body; it is likewise a question of guaranteeing that every understudy has an equivalent chance to create.

Obstructions on the Way to Inclusivity, While the possibility of an additional comprehensive and various world is uplifting, the street to arriving isn't without troubles. Well established inclinations, both cognizant and oblivious, can obstruct progress. Indeed, even in benevolent societies, generalizations and biases can penetrate dynamic cycles and sustain disparity.

Inconsistent admittance to mentorship and movement open doors in the work environment can smother the development of minority people. Oblivious bias in recruiting and advancement cycles can prompt homogeneous administration, proceeding with the cycle. In the event that latest things proceed, it will require around 99.5 years to arrive at orientation equality in the work environment, as per a World Financial Gathering examination.

In training, disparities in subsidizing among well-off and oppressed schools produce inconsistent learning conditions. Understudies from burdened families defy additional obstacles, like an absence of socially reasonable educational plans and restricted admittance to innovation. These obstructions can significantly affect their learning results and future possibilities.

The Weep for Help: People, people group, associations, and state run administrations should all cooperate to advance inclusivity and variety. It's a mobilizing cry to defy our biases, question business as usual, and pursue an additional fair and equivalent world.

1. **Training and Public Mindfulness:** Directing instructive projects that bring issues to light of oblivious inclinations and their consequences is basic. People are bound to put forth purposeful attempts to battle biases when they handle the nuanced ways they influence their insights.

2. **Comprehensive Strategies**: Associations and establishments should have decides that energize incorporation at all levels. This incorporates endeavors to guarantee fair employing, equivalent compensation, and a

chance for headway. Zero-resilience separation strategies pass on a reasonable message that inclusivity is non-debatable.

3. Portrayal and Strengthening: Advancing portrayal at all levels is basic. Administrative roles ought to address the local area's or alternately labor force's variety. At the point when individuals who seem as though them are in, strategic, influential places, it sends a strong message that their fantasies are doable.

4. Establishing Comprehensive Conditions: It is basic to make safe regions where individuals can communicate their concerns and encounters. Comprehensive conversation advances gaining according to each other's perspectives and assists with creating scaffolds of understanding.

5. Educational plan Advancement: Educational programs in instructive establishments ought to be rethought to guarantee that they reflect shifted authentic stories, societies, and viewpoints. This helps understudies in fostering a balanced point of view of the world and difficulties predispositions since early on.

6. Organizations for Joint effort: Legislatures, non-administrative associations, and the corporate area can cooperate to make projects that advance inclusivity and variety. Grant drives, mentorship potential open doors, and local area outreach endeavors can all assist avoided bunches with evening the odds.

7. Nonstop Assessment: Survey and reevaluate strategies and cycles consistently to find regions for development. Inclusivity is a long lasting responsibility that requires adaptability and normal self-reflection.

Inclusivity and variety are more than just popular expressions; they are the groundworks of a fair and evenhanded society. We clear the street for equivalent open doors that permit each person to accomplish their maximum capacity by perceiving the worth of variety and effectively cultivating inclusivity. The way isn't without hardships, however the prize — a world wherein everybody's voice is heard and esteemed — is definitely worth the work. Allow us to recollect that by praising our variety, we open the potential for a more brilliant, more agreeable future for everyone

3.2 Civil rights: Pathways to Value and Incorporation

Civil rights is the brilliant string that ties together the different structure holding the system together in the woven artwork of humankind. It is the firm thought that each individual, paying little mind to foundation, is qualified for equivalent admittance to freedoms, potential open doors, and respect. Civil rights is the base whereupon amicable networks are fabricated — an energizing cry to eliminate treacheries and disparities that stay like breaks in progress' establishment. This significant undertaking is about something other than decency; it is tied in with making a world in which each voice is heard, each want is understood, and each individual is valued.

•Investigating the Separation

Consider a world wherein admittance to school, medical care, and not set in stone by financial status, variety, orientation, or other main qualities. This is the ruthless reality of financial disparities — a gorge that runs profound into our social orders, sustaining patterns of difficulty and prohibition. Perceiving these inconsistencies and facing the inclinations that keep up with them is the most important move toward civil rights.

•Past Deception: Correspondence

Genuine civil rights is in excess of an expression; a pledge to eliminating fundamental obstructions keep equivalent freedoms from being understood. It requires defying laid out biases and reexamining systems that advance unfairness. It isn't just a question of treating everybody the equivalent; it involves perceiving and rectifying verifiable and primary disparities to guarantee that everybody has an equivalent opportunity at progress.

•The Inconsistencies Range

Civil rights includes a large number of worries that influence all parts of life. From orientation pay differences to racial segregation, and instructive disparities to medical services incongruities, the quest for civil rights is a diverse undertaking that requires a total system to address lopsided characteristics and make a more comprehensive society.

•Racial Value: Separating Hindrances

Consider a world where skin tone doesn't restrict one's chances or encounters. Racial equity expects to break the obligations of institutional bigotry by giving underestimated populaces equivalent admittance to training, position, lodging, and equity. A development advocates for the enhancement of recently curbed voices as well as the expulsion of components that propagate separation

•Orientation Equity: Enabling Everybody

Consider a world where orientation doesn't restrict potential. Orientation equity is the objective of equity for all sexes by eliminating inclinations and assumptions that oblige individuals to endorsed jobs. It is tied in with pushing for equivalent compensation, regenerative freedoms, and brutality counteraction, and laying out an air in which everybody, paying little heed to orientation personality, may flourish.

•Shutting the Open door Hole in Financial Consideration

Shut your eyes and envision an existence where financial open door isn't simply an unrealistic fantasy yet a reality for everybody. Monetary incorporation is the work to ensure that individuals from all foundations approach top notch instruction, preparing, and open positions. It is tied in with shutting the financial hole, empowering business, and making a climate in which any ability might flourish.

•Instructive Equity: A Brilliant Future

Think about a general public in which each kid, paying little mind to financial status, approaches a top notch training. Instructive equity tries to even the odds by taking out holes in admittance to instructive assets for oppressed networks. It is a vow to give understudies information, abilities, and open doors that go past their conditions.

•Medical care Value: Recuperating Without Separation

Medical services ought to be a basic liberty, not an honor. Medical care value looks to kill holes in clinical benefit and therapy access. It intends to separate obstructions that keep oppressed bunches from getting to sufficient treatment, guaranteeing that nobody's wellbeing endures because of their conditions.

•Civil rights in real life: Full Developments

History is covered with occasions of people and networks facing unfairness, from social equality missions to LGBTQ+ promotion. These developments started change by addressing severe standards and cleared the way for a more equivalent and comprehensive society. They advise us that civil rights is certainly not a latent pursuit; a source of inspiration requires undeterred purpose.

•The Job of Compassion in Spanning Partitions

Compassion is the capacity to place oneself in the shoes of another, to appreciate their difficulties, and to advocate for their privileges as energetically as our own. Spanning holes requires tuning in, learning, and enhancing minority voices. It is tied in with recognizing that compassion illuminates strategy, discourse, and activity in an equitable society.

•The Force of Activism: Pushing for Change

Activism is the soul of civil rights, a main thrust that moves people and networks to stand firm against foul play. It's a strong instrument for enhancing voices and pushing for foundational change. From road fights to web drives, activism focuses a light on subjects that are much of the time covered in the shadows. It fills in as an update that civil rights is something beyond a hypothetical idea; a development requests cooperative activity to achieve substantial change.

•Instruction as an Adjuster: Mind Strengthening

By furnishing minds with information and decisive reasoning skills, schooling holds the way to taking apart inconsistencies. It's a successful method for separating generalizations, inclinations, and biased ways of behaving. We make the way for a future in which each individual might contribute definitively to society and break liberated from the chains of underlying imbalance by giving extraordinary training to all, no matter what their experience.

•Strategy Pathways for Value Regulation

Regulation is pivotal in laying out the scene of civil rights. It has the capacity to systematize uniformity, safeguard freedoms, and review verifiable wrongs. Legitimate systems, going from hostile to separation regulation to governmental policy regarding minorities in society programs, can make ready for a more equivalent society, guaranteeing that people are dealt with decently as well as have the chance to succeed.

•Enhancement of Voices: The Media and Portrayal

The media can amplify unheard voices and put light on untold stories, making it an amazing asset for civil rights. Portrayal in the media is significant in light of the fact that it challenges assumptions, expands viewpoints, and cultivates sympathy. We advance a more comprehensive culture through enhancing accounts, which perceive the intricacy of human encounters and assurance that everybody's story is told.

•Embracing Intricacy Through Interconnection

Interconnection emerges from the reliance of social personalities like race, orientation, class, sexuality, and others. Perceiving the intricacies of these convergences is basic for handling the exceptional issues that individuals who have a place with various minimized bunches face.

Multifacetedness advises us that civil rights can't be accomplished in a one-size-fits-all way; it requires an exhaustive handle of the different degrees of human experience.

***Making a Superior Future**

As we close to the furthest limit of our examination concerning civil rights, we comprehend that it has been an excursion loaded up with difficulties and victories, disappointments and progress. It is a devotion to making a world wherein the shade of one's skin, orientation personality, or birth conditions don't decide one's destiny. Civil rights is in excess of a fantasy; it is a recognition for our normal humankind and our relentless confidence in the force of progress. By embracing civil rights, we prepare for a future wherein each individual can flourish, each local area can succeed, and each side of the planet can sing an agreeable tune of value and incorporation.

3.3. Advancing Cooperation and Coordinated effort Locally

The force of local area couldn't possibly be more significant in a world that lives on interconnectedness. Networks are the pulsating hearts of society, where individuals get together to share their accounts and structure bonds that cross boundaries. Local area contribution fills in as the magic that binds these ties, empowering support and joint effort that drives positive change and mutual development.

✓ The Significance of Local area

Networks, similar to gardens, require consideration, sustaining, and a cooperative work to flourish. Networks are in a general sense groupings of individuals who share normal interests, values, or topographical spots. They may be basically as little as an area or as extensive as an overall organization connected by a shared objective. The magnificence of networks lives in their variety, with individuals from all beginnings, gifts, and perspectives meeting up to frame a rich embroidery of encounters.

Networks are where we get support, trade thoughts, praise achievements, and work together to take care of issues. They are the spots where our different voices meet up to frame a chorale, amplifying our impact. A dynamic local area encourages a feeling of having a place, which is essential for both individual prosperity and cultural concordance.

✓ The Impact of Local area Commitment

Local area commitment, instead of local area arrangement, is the purposeful course of integrating people into the life and advancement of their local area. It stretches out past enrollment to incorporate dynamic commitment and coordinated effort. At the point when individuals from the local area are involved, they become partners, co-makers, and bosses of positive change.

People are enabled by local area support when they have something to do with choices that impact their lives. It cultivates a feeling of pride and obligation by overcoming any issues between the public authority, associations, and individuals. Drawn in networks are bound to resolve neighborhood issues, foster novel arrangements, and work on broad personal satisfaction.

✓ The Benefits of Local area Association

Reinforced Social Texture: Taking part in local area exercises advances connections, diminishes forlornness, and encourages areas of strength for an of having a place. People are bound to contribute helpfully to their local area when they feel a piece of a more extensive entirety.

•**Informed Independent direction**: Connected with networks pursue informed choices. Individuals get bits of knowledge into complex circumstances when they take part in conversations and discussions, permitting them to pursue informed choices.

•**Nearby Critical thinking**: Who realizes a local area's requirements better than its individuals? Drawn in networks are better ready to perceive and resolve neighborhood issues, bringing about modified arrangements.

Commitment furnishes people with office over their life. At the point when they see their thoughts being embraced and their voices being recognized, they feel engaged.

Various viewpoints invigorate development. At the point when individuals from various foundations cooperate, they offer their viewpoints that might be of some value, bringing about clever fixes and development.

•**Urban Commitment**: Drawn in networks are more dynamic in urban cycles. They vote, advocate for change, and add to the general working of society.

✓ **Local area Commitment Obstructions**

While the advantages of local area commitment are self-evident, there are sure boundaries to investment. These impediments should be distinguished and addressed for commitment exercises to find success.

Absence of Mindfulness: Certain individuals might know nothing about the possibilities of commitment or the expected effect of their cooperation.

•**Time Requirements:** Occupied timetables can make it hard for individuals to commit time to local area exercises.

obstacles to Powerful Correspondence and commitment: Networks are frequently shifted, and language or social contrasts can make obstacles to compelling correspondence and commitment.

✓ **Absence of Trust**: Local area individuals might be reluctant to take part in the event that they have little to no faith in the thought processes of coordinators or foundations.

Networks with restricted assets might find it challenging to construct open stages for contribution.

✓ **Compelling People group Commitment Systems**

Lay out clear and available correspondence courses to keep local area individuals informed about occasions, choices, and open doors for association.

Make a comprehensive working environment where fluctuated assessments are acknowledged and appreciated to guarantee that all voices are heard.

Perceive that different populaces might require different commitment methods. Approaches that are customized to the individual can help with conquering deterrents.

Joint effort with Existing Gatherings: Structure partnerships with neighborhood gatherings, schools, and companies to exploit existing organizations and assets.

✓**Limit Building**: Give preparing and studios to assist local area individuals with working on their abilities and certainty, permitting them to offer all the more actually.

Perceive and Observe Commitments: Perceive and commend crafted by dynamic local area individuals. This affirmation can act as a consolation to take part.

✓**Innovation and Local area Support**

The computerized unrest has modified the scene of local area collaboration. Web-based entertainment, online gatherings, and virtual municipal centers have expanded commitment exercises' arrive at past topographical limits. Innovation considers quick correspondence, permitting gatherings to associate and team up continuously. Nonetheless, it is basic to keep up with comprehensive and available advanced contribution for all, tending to the computerized partition that can exist in certain areas.

✓ **Commitment's Expanding influence**

At the point when a solitary stone is dropped into a lake, swells spread outward and contact the whole surface. Likewise, people group inclusion has an expansive effect that stretches out past the prompt members. People who are involved become change representatives, motivating others to reach out and contribute their abilities and viewpoints.

As additional individuals take an interest, there is a more noteworthy feeling of local area possession and obligation. This prompts more viable critical thinking, better navigation, and a common obligation to the prosperity, all things considered.

✓ Keeping up with Local area Commitment for Long haul Effect

Believe a local area to be an ensemble, with every part playing an alternate instrument. Each instrument should be tuned and each performer should be receptive to the aggregate cadence for the ensemble to reverberate really. Additionally, keeping up with local area commitment requires progressing exertion, care, and transformation to guarantee that shared congruity endures.

✓ Making a Commitment Culture

Fostering a commitment culture starts with recognizing that local area individuals are not aloof beneficiaries but rather dynamic members in impacting their normal climate. It's tied in with valuing each info, regardless of how huge or small, and perceiving that various perspectives enhance the discussion.

Local area pioneers ought to give ordinary open doors to interest to support commitment. Steady connection, whether through municipal events, studios, or online stages, makes all the difference for the discourse. Besides, offering thanks for commitments, even through straightforward motions, for example, cards to say thanks or public affirmations, urges individuals to remain involved.

✓ **Schooling for Strengthening**

Schooling is a basic part of commitment maintainability. Local area individuals are bound to remain committed when they figure out the effect of their action. Studios that make sense of how choices are made, the job of neighborhood government, and the consequences of different choices empower individuals to take an interest all the more really.

By giving admittance to information, local area pioneers furnish people with the devices expected to explore troublesome difficulties. This cultivates trust as well as better instructed and helpful discussion.

✓ From Investment to Proprietorship

Supported people group commitment develops into possession, as local area individuals become stewards of their common surroundings. This stage is featured by a move from taking part in occasions and discussions to effectively driving exercises. People who take possession are more put resources into the results, it are not to no end to guarantee that endeavors.

Also, proprietorship energizes intergenerational congruity. At the point when more youthful individuals see their older folks and companions partaking, they are bound to proceed with the custom. Passing down the inclusion light ensures that the exuberance of the local area is kept up with.

✓ Utilizing Innovation Dependably

Nowadays of quick innovation advancement, using computerized innovations can significantly further develop commitment endeavors. Web-based entertainment, online reviews, and virtual occasions give cooperation valuable open doors that cross actual boundaries. In any case, innovation ought to be drawn nearer cleverly, guaranteeing that it increments inclusivity as opposed to irritating imbalance.

Endeavors should be made to connect the computerized partition by giving access and preparing to the individuals who may be new to these apparatuses. Additionally, an equilibrium ought to be kept up with among computerized and eye to eye contribution to keep up with the glow of individual ties.

✓ **Influence Appraisal and Methodology Adjustment**

An information driven procedure is expected for supportability. Estimating the effect of commitment exercises routinely empowers the disclosure of fruitful strategies as well as regions for development. Support rates and subjective criticism, for instance, could give knowledge into the adequacy of commitment endeavors.

Flexibility is fundamental. As people group change, so do their requirements and inclinations. Pioneers should change commitment strategies in light of evolving conditions, new advancements, and creating concerns. Commitment stays important and engaging with a versatile methodology.

✓ Defeating Manageability Deterrents

While the advantages of progressing local area commitment are colossal, issues that imperil its continuation can create.

1. Fatigue: Assuming that individuals feel overburdened, long haul commitment can prompt burnout. This issue can be reduced by turning influential positions and empowering collaboration.

2. Segment Movements: Networks go through segment shifts over the long haul. It is basic to guarantee that commitment exercises are comprehensive and receptive to evolving socioeconomics.

3. Financial, political, or social movements could avoid concentration and assets from commitment endeavors. During such occasions, local area versatility is scrutinized, and savvy fixes are expected to keep individuals locked in.

4. Keeping up with Force: There is a gamble of carelessness after first achievement. Keeping up with force is made more straightforward by reliably conveying the significance of commitment and the effect of collective endeavors.

Local area contribution is a ceaseless obligation to empowering interest and cooperation as opposed to a one-time occasion. It is tied in with perceiving that the strength of a local area is gotten from the dynamic support of its kin. Economical commitment gives a gradually expanding influence of good change that broadens well past the quick members, from developing a culture of support to enabling through instruction and innovation.

As we weave the texture of dynamic networks, remember that a vivacious local area, similar to a developing nursery, requires normal consideration. We guarantee that the orchestra of our consolidated endeavors proceeds to fit and create an enduring effect for people in the future by empowering cooperation, encouraging joint effort, and acclimating to evolving conditions.

3.4. Essential Requirements Access: Guaranteeing Wellbeing, Schooling, and Prosperity

The strings of wellbeing, schooling, and prosperity are fundamental in the wide texture of common liberties. They weave the fragile texture that advances individual potential and cultural progression. Admittance to these essential necessities is a basic right that opens the way for a more splendid, all the more fair future. Giving admittance to wellbeing, instruction, and prosperity is an ethical obligation that lays the right foundation for a fruitful society from the support to the homeroom, from the specialist's office to the public venue.

✓ Health: An Establishment for Success

Consider a world where everybody has the potential chance to awaken to a sound morning and the danger of preventable sicknesses doesn't loom over their heads. Admittance to satisfactory medical services is fundamental for life as well as for strengthening and achievement.

Individuals can work, learn, and take part completely in the public arena when they approach essential medical care administrations. Convenient mediations can hold gentle diseases back from deteriorating into serious wellbeing emergencies, permitting individuals to carry on with useful lives liberated from avoidable agony.

Besides, wellbeing is a point of support that advances instructive fulfillment. A sound youth is bound to go to class consistently, retain data all the more effectively, and succeed in their examinations. By putting resources into open medical services, we lay the foundation for a drawn out pattern of further developed wellbeing, training, and prosperity.

✓ Education: Enlightening Advancement Courses

Training is the beacon that guides us from obliviousness to edification, from prohibition to strengthening. The gift gives individuals the devices they need to construct a superior life for them as well as their networks. Training, then again, is tied in with developing a culture of interest and decisive reasoning, not simply study halls and course readings.

At the point when each youngster, paying little heed to foundation, approaches great schooling, we release their capability to be change specialists. Instruction crosses borders, taking into consideration the trading of groundbreaking thoughts, advancements, and joint efforts that impel progress. A power empowers individuals to break the pattern of neediness, challenge cultural standards, and dream past their conditions.

We plant the seeds of future pioneers, masterminds, and issue solvers by guaranteeing evenhanded admittance to instruction. Schooling accomplishes something beyond fill minds; it likewise powers minds and engages individuals to shape a world that mirrors their objectives.

✓ The Human Soul and Prosperity

Prosperity is more than essentially actual wellbeing; it is tied in with supporting a person's comprehensive necessities for them to flourish inwardly, intellectually, and socially. Prosperity isn't an extravagance; it is the center of a masterfully carried out life.

Prosperity expects admittance to emotional well-being assets, places of refuge, and a steady local area. People are more prepared to navigate life's intricacies when they have the apparatuses to oversee pressure, adapt to issues, and look for help when required.

Putting resources into one's prosperity involves valuing the innate worth of recreation, amusement, and social advancement. These parts of life are not paltry; they add to a solid and satisfied life. Admittance to public venues, outside regions, and far-reaching developments enhances individuals' lives through building connections and a feeling of having a place, the two of which are fundamental for human thriving.

✓ Bringing Down Boundaries to Access

The way of guaranteeing admittance to wellbeing, schooling, and prosperity isn't without troubles. Many individuals experience obstructions that keep them from practicing their key freedoms.

1. Financial Inconsistencies: Destitution regularly makes access obstacles. People with little funds might experience issues bearing the cost of medical care, superb schooling, or exercises that advance prosperity.

2. Geographic Confinement: Distant people group might need admittance to significant administrations, including medical care offices and instructive foundations. This detachment could demolish existing awkward nature.

3. Orientation Disparity: Unfair practices and customs could restrict admittance to unmistakable sexual orientations. Guaranteeing equivalent access for everything is significant for progress.

4. Disgrace and Mindfulness: Psychological well-being troubles oftentimes convey a shame that ruins people from getting care. Bringing issues to light and supporting open talk is vital to beating this deterrent.

5. Social Responsiveness: Medical services and schooling should be adjusted to regard social standards and language assortment. A one-size-fits-all approach might distance explicit societies.

✓ **Making a Comprehensive Future**

Admittance to wellbeing, schooling, and prosperity isn't simply an issue of strategy; it's an issue of mankind. It's tied in with perceiving the natural worth of each and every human and guaranteeing that their essential requirements are fulfilled, no matter what their conditions. It's tied in with laying out a general public where everybody has the chance to realize their maximum capacity and add to the aggregate advancement.

Legislatures, associations, and people all play a part to play in this endeavor. Strategies that focus on admittance to medical care and schooling, as well as measures that advance prosperity, are crucial stages toward a more comprehensive future. Coordinated efforts between general society and business areas can build the effect of these endeavors and guarantee that nobody falls behind.

As we walk forward, let us recall that admittance to wellbeing, instruction, and prosperity isn't an honor to be acquired; it's an option to be protected. Winding around a general public where the strings of wellbeing, training, and prosperity construct an embroidery of versatility and progress that encompasses all of us by cooperating to separate boundaries, span holes, and advance value.

✓ Individual Strengthening, People group Improvement

The quest for admittance to wellbeing, instruction, and prosperity isn't simply a reason; a development can reshape civic establishments from the grassroots up. People who are enabled to deal with their wellbeing, training, and prosperity have a broad effect that waves through families, networks, and even countries.

✓ Information Based Strengthening

Information is the way to making the way for strengthening. Giving individuals precise data about their wellbeing, instructive potential outcomes, and ways of further developing their prosperity enables them to pursue informed choices. People are enabled to effectively deal with their wellbeing through wellbeing training that shows anticipation, solid ways of behaving, and early intercession.

Essentially, instruction that goes past repetition learning and supports decisive reasoning enables individuals to battle imbalances and supporter for their freedoms. Individuals are better situated to get quality medical care and schooling when they grasp their freedoms and know about accessible assets.

Advancing prosperity additionally involves showing the capacity to appreciate anyone on a profound level and versatility capacities. Showing individuals how to deal with pressure, look for help, and fabricate survival strategies works on their ability to manage life's deterrents.

✓ **Support Locally**:

Admittance to wellbeing, instruction, and prosperity isn't simply a singular excursion; it is a local area exertion. Networks go about as wellbeing nets, getting people who fall and giving a stage to joint effort and help. Local area association is basic in guaranteeing that everybody approaches these major necessities.

The effect is gigantic when networks rally to push for further developed medical care offices, available training, and safe sporting spaces. Drawn in networks are bound to consider establishments responsible, advance evenhanded asset dispersion, and encourage a climate wherein everybody's requirements are considered.

Local area drove projects additionally can possibly resolve explicit issues that bigger gatherings might disregard. From neighborhood wellbeing fairs to education programs, these ventures address the local area's particular necessities, it is abandoned to guarantee that nobody.

✓ Intergenerational Cycle Breaking

The capacity to disturb intergenerational patterns of neediness and hindrance is one of the best pieces of guaranteeing admittance to wellbeing, training, and prosperity. Kids' life ways change considerably when they experience childhood in an air where medical care is accessible, schooling is esteemed, and prosperity is focused on.

Youngsters who approach excellent training are bound to break out from the pattern of low proficiency and restricted vocation possibilities. Further developed medical care administrations guarantee that preventable sicknesses don't obstruct youngsters' development and improvement. Moreover, a prosperity culture trains them to stress taking care of oneself and profound flexibility, laying the basis for a more joyful future.

This change affects the whole local area, in addition to the close family. As these enabled people mature into contributing citizenry, they act as good examples for other people. A chain response is gotten rolling after some time, steadily rescuing whole networks once again from destitution and creating a more prosperous society.

✓ **An Ethical constraint and a Common Obligation**

Admittance to wellbeing, training, and prosperity is an ethical need as well as a major basic freedom. As a worldwide local area, we are committed to guarantee that each individual, paying little mind to situation, has the valuable chance to carry on with a sound, taught, and satisfied life.

States, organizations, organizations, and people all offer this obligation. A revitalizing cry cuts across nations and thoughts. At the point when we cooperate to eliminate snags, span holes, and advance value, we construct a world in which each kid might dream unbounded, each individual approaches significant administrations, and each local area flourishes.

We find the center of mankind in this joint exertion — the acknowledgment that our fates are entwined and our progression is related. Winding around an embroidery of trust, versatility, and shared flourishing that wraps the whole planet through supporting admittance to wellbeing, schooling, and prosperity. As we push ahead, remember that by guaranteeing that everybody's essential requirements are tended to, we lay the basis for a super

Chapter 4

Financial Reasonability: Accomplishing Success

4.1 Round Economy: Changing from Direct to Maintainable Plans of action

Consider a future wherein squander doesn't gather in landfills, assets are perpetually recovered, and financial extension doesn't come to the detriment of natural harm. This vision lies at the core of the round economy, a progressive idea that challenges the customary straight "take, make, arrange" model and offers a street to long haul achievement. In a world adapting to serious ecological worries, it is certainly not a decision to embrace the roundabout economy; it is a necessity for preserving our planet and guaranteeing a prosperous future for people in the future.

Rethinking Progress

The straight economy that has long overwhelmed our worldwide monetary scene is set apart by an upsetting reality: limited assets are mined, transformed into items, consumed, and eventually disposed of as trash. This straight technique has sped up the exhaustion of normal assets, contamination, and a developing waste situation. It is an unreasonable and eventually reckless way.

The roundabout economy, then again, challenges this worldview by rethinking progress. Its center objective is to separate from financial development from asset utilization and ecological corruption. It proposes a framework wherein assets are utilized however long practical and squander is arranged out of the situation.

Shutting the Hole

The roundabout economy isn't an unrealistic fantasy; it is a reasonable arrangement in view of imaginative reasoning and practical ways of behaving. One of its key standards is "shutting the circle," which infers planning merchandise and frameworks that permit materials to be recuperated and reused. Rather than winding up in landfills, things are intended to be dismantled, remanufactured, or reused, broadening their lives and limiting the requirement for virgin assets.

For instance, the style area, which is known for its inefficiency, is embracing round strategies. Brands are exploring different avenues regarding garments rental administrations, utilizing reused textures, and planning items for simple dismantling. This technique diminishes the business' natural impact as well as produces new financial possibilities through inventive plans of action.

From possession to get to

The idea of possession changes in the roundabout economy. People and organizations can get to item usefulness as administrations as opposed to claiming things through and through. This change from proprietorship to get to cultivates longer item lives as well as boosts makers to plan things in view of solidness and repairability.

Consider approaching the advantages of a clothes washer without the weight of proprietorship. This strategy pushes producers to assemble superior grade, enduring machines that are effortlessly kept up with and moved along. It's a mutually beneficial arrangement: buyers get more straightforward admittance to things, while makers make items with longer lifecycles.

Embracing Advancement and Cooperation

The round economy depends on cross-industry development and coordinated effort. It energizes architects, specialists, and business visionaries to go past the container and rethink how items are produced, dispersed, and consumed. It additionally advances joint effort among different partners, going from legislatures and ventures to shoppers and non-administrative associations.

3D printing, which permits things to be made on request and with negligible waste, is changing areas. Joint efforts among organizations and nearby networks could bring about asset sharing activities in which extra materials from one activity become inputs for another. This recoveries junk as well as cultivates a harmonious cooperation among organizations and their current circumstance.

Financial Advantages and Strength

In opposition to prevalent thinking, the round economy has huge financial advantages. Organizations can cut costs and lift productivity by diminishing asset use and waste result. Besides, the round economy can give new income transfers through item as-a-administration models, fix administrations, and the selling of repaired products.

Besides, the roundabout economy works on financial versatility. As worldwide inventory networks are disturbed and asset shortage turns out to be more intense, associations that have embraced roundabout strategies are better situated to meet these issues. They depend less on virgin assets and are better ready to answer changing economic situations.

Embracing the roundabout economy involves criticalness even with gigantic natural emergencies. The direct model has taken us to a position where the exhaustion of normal assets and the collection of waste imperils the very frameworks that support life on The planet.

Legislatures can assume a part by ordering guidelines that energize roundabout practices, for example, expanded maker obligation and tax cuts for asset proficient items. Organizations can start to lead the pack by reconsidering their products, carrying out reasonable inventory network processes, and helping out different partners to close the circle.

Shoppers have options also. People can assist with creating interest for a roundabout economy by supporting firms that esteem roundabout works on, requesting tough and repairable items, and taking part in item sharing or rental organizations.

A Progress Outline: Changing from a straight to a roundabout economy requires a thorough methodology that incorporates plan, creation, utilization, and waste administration. This change is a bit by bit process that requires responsibility, collaboration, and continuous development.

✓ **1. Plan for Maintainability**: At the core of the roundabout economy is plan. Items ought to be planned in light of the end client, with thought given to how promptly they might be dismantled, fixed, and reused. Fashioners assume a basic part in reevaluating materials, choosing sustainable assets, and diminishing unsafe parts. The idea of "support to support" plan ensures that items are made to be regenerative instead of exhausting.

✓ **2. Embrace Sharing Stages**: The ascent of the sharing economy gives an amazing opportunity to integrate roundabout ways of behaving into day to day existence.

Sharing stages for all that from instruments and attire to vehicles and office spaces empower asset sharing, dragging out item lifecycles and diminishing the requirement for superfluous possession. Cooperative utilization techniques are impeccably lined up with the roundabout idea.

✓ **3.** **Reconsidering Utilization**: Our choices as customers are significant. We can put forth a deliberate attempt to help items that are solid and be fixed. Picking revamped or utilized things over pristine ones limits interest for virgin assets. We can likewise decide to participate in reclaim programs given by firms to guarantee proper removal toward the finish of an item's life.

✓ **4. Empowering Fix and Updates:** A vital principle of the round economy is the ability to fix and overhaul things as opposed to dispose of them. Organizations might help by giving fix administrations, giving substitution parts, and making merchandise that are not difficult to fix. This

change can be advanced by regulation that ensures clients'
on the whole correct to fix.

✓ **5. Progressing Supply Chains**: A round economy
needs reexamining old inventory network ideal models.
Organizations can source materials locally, structure asset
sharing associations, and select providers who stick to
round standards. This move limits the ecological effect of
transportation as well as supports the neighborhood
economy.

✓ **6. Instructing and Upholding**: Bringing issues to light
about the roundabout economy and its advantages is basic.
Teaching purchasers about their part in expanding interest
for round items and administrations permits them to settle
on additional educated choices. Backing endeavors can
likewise impact official changes that support roundabout
practices and beat inefficient ones down.

✓ **7. Strategy and Guideline:** States play a basic part in
hurrying the progress to a round economy. Arrangements
that advance maintainable practices, like broadened maker

obligation and eco-naming, assist with evening the odds for roundabout firms. Tax cuts for reusing and remanufacturing can support utilization significantly further.

✓ **8. Joint effort and Advancement:** Cooperative endeavors between undertakings, legislatures, non-administrative associations, and scholastic foundations are basic for propelling the round economy. Associations can advance the trading of best practices, co-make effective fixes, and address fundamental impediments that hinder development.

✓ **9. Estimating Effect:** As the roundabout economy gains forward movement, evaluating its impact is basic. Measurements that go past ordinary monetary measures, for example, how much garbage diminished, materials moderated, and emanations stayed away from, give a full evaluation of improvement.

A Joint Commitment

The transition to a round economy requires a cooperative exertion that crosses lines, ventures, and philosophies. An errand requests solid initiative, innovative reasoning, and a commitment to safeguarding our reality for present and people in the future.

This change is about something beyond decreasing waste; it is additionally about reclassifying esteem. It is tied in with tolerating that financial advancement and ecological stewardship can coincide. It is tied in with rethinking outcome as far as cleverness and recovery as opposed to extraction and exhaustion.

The roundabout economy is in excess of a hypothesis; it is a guide for long haul success. It gives a street forward that disappears from the edge of ecological catastrophe and toward a future where progress is related with prosperity. It's a push to reexamine corporate strategies, customer inclinations, and regulative structures. By taking on the round economy, we put into high gear a transformation with the possibility to reexamine ventures, change economies, and recuperate our planet.

We honor the reliance of all life on Earth as we go from direct to round. We perceive that our assets are restricted, however our commitment is boundless. We change our point of view from one of utilization to one of stewardship. The roundabout economy isn't just a decision; it's a heritage we leave for people in the future — a tradition of maintainability, strength, and the getting through magnificence of our common world.

4.2 **Dependable Utilization:** Advancing Moral and Neighborhood Practices

In a world brimming with choices, each buy we have makes a tremendous difference — on the climate, on society, and our aggregate fate. Dependable industrialism is in excess of a trademark; a directing way of thinking enables people to settle on decisions that line with their qualities and add to everyone's benefit. By advancing moral and neighborhood ways of behaving, we might move our purchasing behaviors from thoughtless utilization to careful support, guaranteeing a more manageable and fair society for all.

✓ The Impact of Our Decisions

Each time we open our wallets, we vote in favor of the sort of world we need to live in. The things we pick, the organizations we support, and the ways of behaving we support send an assertion to areas and economies. Dependable utilization perceives this potential and outfits it to impact positive change.

Moral and neighborhood ways of behaving are established on a significant enthusiasm for our worldwide local area's association. They comprehend that behind each item is a convoluted trap of living souls and natural assets. We can drive interest for items that focus on individuals and the climate over benefit alone by settling on careful decisions

✓ Moral Utilization: An Ethical Compass

Moral commercialization involves picking choices that regard basic freedoms, energize fair work norms, and limit ecological damage. It is tied in with helping firms that keep moral guidelines all through their stockpile chains, from natural substance acquisition to assembling and conveyance.

At the point when we buy things that are confirmed as fair exchange, mercilessness free, or environmentally cordial, we hoist the voices of underestimated laborers, save biological systems, and send a reasonable explanation that shifty strategies are not satisfactory. We stay standing for a general public where empathy and manageability are non-debatable beliefs by declining to help businesses that benefit from human torment or environmental calamity.

✓ Nearby Utilization: Building People group

Nearby utilization is inseparable from capable utilization. We foster unique networks, make occupations, and diminish the carbon impression of significant distance transportation by supporting nearby organizations. Neighborhood makers habitually esteem quality above amount, and their exercises are inseparably connected to the wellbeing and prosperity of their networks.

At the point when we purchase items from nearby ranchers' business sectors, we appreciate new, occasional food varieties, yet we additionally help to safeguard neighborhood rural practices. Additionally, buying merchandise from neighborhood craftsmans and specialists assists with keeping up with social heritage and workmanship while reducing the ecological effect of large scale manufacturing.

✓ The Benefits of Dependable Utilization

Natural Stewardship: Capable commercialization assists with keeping up with environments, forestall contamination, and decrease squander. We help to save normal assets and moderate environmental change by advancing eco-accommodating items and practices.

Human Nobility: Moral utilization guarantees that the things we purchase are not spoiled by double-dealing, constrained work, or common freedoms infringement. It advances laborers' privileges, fair pay rates, and safe working circumstances, bringing about an additional equitable and evenhanded world.

Nearby buying helps nearby economies, giving position and supporting neighborhood life. It fosters a feeling of local area and shared liability regarding the prosperity, everything being equal.

✓**Higher expectations no matter** what: Dependable commercialization trains us to underscore better standards without ever compromising. By putting resources into all around made, enduring items, we diminish the requirement for regular substitutions, eventually setting aside cash and assets.

✓**Catalyzing Change:** As customers request more moral and ecological items, ventures answer by rethinking their techniques. This has a flowing impact that prompts foundational change, compelling organizations to embrace more dependable practices.

✓ Overseeing Troubles

While dependable utilization enjoys various benefits, it isn't without challenges. The fascination of modest, efficiently manufactured merchandise can be alluring, and deception can make it hard to recognize genuinely moral and nearby items from those that seem, by all accounts, to be. Buyers can deal with these issues and pick decisions that mirror their qualities with expanded information, training, and exploration.

✓ Teaching Ourselves As well as other people

Instruction is the most vital move towards enabling moral utilization. We become educated shoppers who can settle on choices lined up with our qualities by finding out about supply chains, figuring out accreditations, and concentrating on the acts of the organizations we support.

Moreover, as mindful buyers, we can teach others. Sharing data on moral and neighborhood works on, discussing the results of our decisions, and pushing for change in our networks can create a wave outcome of mindfulness and activity.

✓ Strategy and Industry Changes

Dependable utilization isn't simply the obligation of individual shoppers; it requires a cooperative exertion traversing enterprises and legislatures. Policymakers can boost moral and neighborhood rehearses by ordering arrangements that reward manageability and put unsafe ones down. Organizations can rehearse genuine revealing, embrace round economy thoughts, and underscore the prosperity of their workers and networks.

Mindful utilization isn't a penance; it is an interest in a superior future. It's a statement that our convictions matter and that each choice we make adds to the embroidery of our common fate. By advancing moral and neighborhood ways of behaving, we make a texture of sympathy, maintainability, and local area versatility that encompasses every one of us.

In a general public where commercialization has broad ramifications, savvy decisions are a light of trust. They mirror the conviction that each buy can shape ventures, drive change, and construct a general public that values the two individuals and the earth. As dependable shoppers, we have the honor and commitment to be insightful stewards of our decisions, leaving a tradition of constructive outcomes for people in the future.

4.3 Green Money: Adjusting Ventures to Supportability Objectives

Green Money: Adjusting Ventures to Supportability Objectives, A critical response creates at the junction of natural emergencies and monetary desires: Green Money. This unique technique rises above conventional monetary limits, introducing another period where ventures and manageability objectives are amicably entwined. As the dangers of environmental change become more clear and the inclination for activity gets stronger, the idea of adjusting ventures to manageability objectives builds up momentum more than ever.

Consider a world where the customary idea of money isn't just about boosting benefits yet in addition about expanding helpful impact. This is the substance of Green Money. At its center, it is tied in with assigning subsidizing to drives that guarantee monetary advantages as well as effectively add to ecological and social prosperity. The change in perspective here is significant: from taking advantage of nature to sustaining it, from transient increases to long haul solidness, and from segregated monetary profits to comprehensive advantages for the two people and the earth.

The significance of this shift can't be accentuated. As ozone depleting substance outflows rise and biodiversity declines, the worldwide local area faces an existential emergency. Green Money enters the scene as a strong superhuman, outfitted with a triple-main concern approach that focuses on individuals, planet, and benefits.

Its impact originates from ability to redirect capital from organizations hurt the climate and society and toward those that embrace environmentally friendly power, supportable horticulture, clean innovation, and different drives.

This advantageous connection among cash and supportability is fastidiously developed through various channels. Green bonds, for instance, act as channels among financial backers and environmentally ideal ventures. These monetary instruments give a reasonable open door to financial backers to help environmental change undertakings like sustainable power establishments, energy-proficient structures, and eco-accommodating foundation. Green bonds draw in the people who need to associate their ventures with their goals because of their straightforwardness and reason driven character.

In any case, the extent of Green Money reaches out past these securities. Right now, monetary organizations are rehashing their jobs. Banks, resource directors, and protection organizations are progressively integrating ecological, social, and administration (ESG) contemplations into their dynamic cycles. This coordination isn't just about appearances; it's tied in with building a strong monetary framework that is receptive to the truth of an impacting world.

Think about the idea of effect venture. It exhibits that monetary returns don't need to be in conflict with long haul yearnings. Influence financial backers forcefully look for open doors that have a decent effect while as yet giving serious monetary returns. This double design is a perfect representation of how Green Money is reevaluating the contributing climate. A trying takeoff from the standard way of thinking benefit and morals are hostile together.

As Green Money acquires pace, in addition to the regular scene benefits. The monetary field will benefit too. Green ventures support work creation, development, and financial flexibility. Consider a future where interests in environmentally friendly power produce occupations in provincial networks, manageable farming strategies increment food security, and clean advancements change whole businesses. This future is reachable, powered by Green Money swells that enhance positive change across enterprises.

Be that as it may, similarly as with each weighty idea, hardships emerge. Cynics might contend that the monetary penances expected to change toward maintainability are excessively perfect. In any case, this is a confined viewpoint that disregards the drawn out advantages of a practical planet. Actually the expenses of ecological corruption, social shamefulness, and environment related calamities much outperform any momentary monetary prizes.

The way to appropriately adjusting ventures to natural objectives requires worldwide coordinated effort. States, administrative organizations, monetary foundations, and common society should cooperate to lay out a biological system that backings and boosts green ventures. This could incorporate the making of normalized ESG measures, tax cuts for manageable endeavors, and rules requiring the revelation of natural risks.

Green Money is, at its center, an ethical movement as opposed to a specialized creation. It is a statement that we can never again dismiss the association of all life on The planet. It is a statement that the prosperity of our planet is inseparably attached to the prosperity of our economies. It's a mindfulness that the world we pass on to people in the future is molded by the choices we make today.

In the expansive texture of presence, Green Money is the string that associates financial aspects and morals. It's a revitalizing cry to redirect the waterways of money toward the nurseries of maintainability. It fills in as an update that our speculations can be something other than numbers on a monetary record; they can be impetuses for positive change. So let us embrace this change in which money turns into a power for good and the objective of wealth fits with the quest for a flourishing planet. The opportunity has arrived to coordinate ventures with manageability objectives, and with every speculation made, we draw nearer to a greener, more promising time to come. Green Money arises as the change-producer in the muddled dance of money and maintainability. It concedes that the customary model of monetary development, which is every now and again powered by the deficiency of regular assets, is as of now not feasible. All things being equal, it advances a sound beat in which monetary extension is adjusted by natural consideration and social consideration.

One of the incredible characteristics of Green Money is its ability to amplify the influence of aggregate activity. By empowering people, associations, and organizations to coordinate their ventures toward maintainable tasks, a gradually expanding influence is made that resounds across the worldwide economy. This unique cooperation between little activities and huge repercussions gives a striking image of how interrelated our reality genuinely is. Think about the job of institutional financial backers in this change. Annuity reserves, sovereign abundance assets, and enrichments handle colossal amounts of cash. At the point when these monetary titans commit devotion to maintainability, they send serious areas of strength for a to the venture world. Their reception of ESG standards and assignment of assets to green resources validate the monetary practicality of feasible financial planning.

The monetary world is changing, moving from a momentary benefit driven direction to a drawn out esteem situated point of view. Green Money speeds up this change by pushing financial backers to go past momentary gains and think about the drawn out effect of their choices. This pattern is predictable with the idea of intergenerational value, which perceives that the assets we consume and the choices we make today have broad ramifications for people in the future.

This change, in any case, isn't exclusively persuaded by consideration. Another flood of financial backers is perceiving the likely monetary awards of lining up with supportability drives. As the worldwide local area fixes its hold on ecological standards, associations that are proactive in carrying out economical practices will be better situated to flourish in a future where fossil fuel byproducts are evaluated and asset shortage ends up being more self-evident.

It is basic to perceive that the shift to Green Money is definitely not a one-size-fits-all activity. It esteems the variety of enterprises, areas, and economies. In emerging nations, for instance, the issue is to adjust the requirement for financial advancement with ecological safeguarding. Green Money could assume a basic part in funding projects that sidestep conventional, asset serious development ways, impelling these countries toward long haul thriving.

Another viewpoint that adds to Green Money's reverberation is mechanical development. From blockchain supporting straightforward inventory chains to man-made brainpower streamlining energy effectiveness, innovation opens up new vistas of potential. The coordination of fintech with manageability gives constant following of ESG execution, democratizes admittance to green speculations, and further develops risk appraisal methodology.

The significance of schooling and mindfulness in this far and wide change couldn't possibly be more significant. As Green Money reshapes the monetary climate, people should have the information and instruments to go with taught choices. The idea of "monetary education" is developing to incorporate realizing monetary business sectors as well as perceiving the moral components of speculation choices.

Cooperation between the general population and business areas is vital for the maintainability of this development.

To energize feasible ventures, states could give regulative motivating forces, for example, tax reductions and appropriations. Administrative specialists can make express structures for detailing ESG measurements, expanding straightforwardness and responsibility. In the interim, organizations can take the risk to advance by creating items and administrations that compare with the rising tide of cognizant commercialization.

In the stupendous account of Green Money, achievement is assessed in incomes, yet additionally in expanded biological system strength, decreased social imbalances, and diminished environment dangers. It's an account of thriving reconsidered, where monetary development meets with natural insurance. It's a guide for a future wherein monetary establishments are engineers of positive change, making structures that last past financial quarters and yearly reports.

As the parts of this novel create, the vision turns out to be clear: Green Money can modify the narrative of our planet's fate. It cuts across lines, convictions, and socioeconomics, communicating in a worldwide language of maintainability and progress. It's a source of inspiration, a call to reroute the stream of capital into productive grounds that bear the prizes of a greener, more libertarian world.

Green Money isn't a prevailing fashion, yet rather a crucial change by they way we see and utilize our monetary assets. It's an orchestra wherein ventures blend with supportability objectives, giving melodies of good change that resound past ages. Green Money makes an embroidery of thriving by interfacing profit with reason. It's a tribute to our normal commitment to safeguard the world and guarantee its abundance for people in the future.

4.4 Advancement and Green Innovations: Driving Financial Development and Progress

Advancement and Green Innovations: Driving Monetary Development and Progress

In the consistently changing range of worldwide worries, one issue stands apart as both a basic concern and a remarkable open door: the union of development and green advancements. As the globe wrestles with ecological debasement, asset consumption, and the impacts of environmental change, the presentation of leading edge green innovation not just offers a brief look at trust for a maintainable future yet additionally readies the way for exceptional monetary development and progress.

Over the entire course of time, advancement has been at the core of human advancement. From the innovation of the wheel to the advancement of the web, each forward-moving step has been pushed by the relentless human soul to tackle issues and work on our personal satisfaction. Today, our joint test is to find novel responses to the convoluted hardships presented by ecological debasement. Green innovations are a gathering of developments, procedures, and applications that plan to decrease the adverse consequence of human movement on the climate.

The significance of tending to environmental change and asset shortage can't be underscored. Green innovations cover many disciplines, from sustainable power sources like sun based, wind, and hydroelectric capacity to energy-proficient structure plans, manageable farming techniques, and waste decrease methodologies. These innovations not just guarantee to diminish the ecological harm brought about by conventional ways yet in addition to deliver a surge of monetary open doors that can drive critical development and headway.

The foundation of new business sectors is one of the main drivers of monetary development connected to green advancement. As legislatures, companies, and people value the significance of maintainability, there is a rising interest for harmless to the ecosystem items and administrations. This developing interest opens up new open doors for undertakings, going from the advancement of electric vehicles and shrewd lattices to eco-accommodating purchaser things. This, thusly, makes work prospects and empowers financial action.

Moreover, the change to green advances can possibly change whole ventures. Think about the energy area: the move from petroleum derivatives to sustainable sources sabotages laid out energy creation designs, bringing about the production of clean energy behemoths. This cuts ozone harming substance emanations as well as supports a rush of development in energy capacity, network control, and proficiency improvement. Generally, the reception of green innovations has a far reaching influence that infests various areas and cultivates the development of new organizations.

The sunlight based energy transformation is a superb representation of this peculiarity. Sunlight based power has advanced from a particular innovation with restricted applications to one of the world's quickest developing wellsprings of environmentally friendly power. Sunlight based charger proficiency forward leaps, joined with falling creation costs, have made sun oriented power a reasonable option in contrast to conventional energy sources. This structural commotion has changed the energy climate as well as created advancement in energy capacity frameworks and matrix combination methodologies.

Moreover, green innovations can change metropolitan preparation and framework improvement. As the total populace movements to urban communities, the need for feasible urbanization becomes basic. Green structure materials, savvy city advances, and productive public travel frameworks are changing metropolitan habitats into eco-accommodating living communities. This change takes care of ecological issues as well as gives a prolific field to monetary development, as metropolitan development communities draw in speculation and ability.

It is imperative to stress that the monetary advantages of green innovations stretch out past quick monetary profits. Countries can further develop their energy security by putting resources into supportability. Reception of green practices can add to better populaces, lower medical services costs, and greater of life all simultaneously. These combined impacts make a positive criticism circle in which financial development and progress are inseparably connected to natural prosperity.

In any case, the way to a supportable future isn't without hindrances. The reception of green advances requires critical interests in exploration, improvement, and foundation. Legislatures, industry, and the scholarly community should cooperate to beat innovative difficulties, smooth out administrative systems, and guarantee that the advantages of green innovation are accessible to all sections of society. Moreover, there might be resistance from laid out businesses that stand to experience the ill effects of the progress to additional reasonable practices. To defeat these hindrances, visionary initiative, key readiness, and a strong obligation to a typical future are required.

The marriage of development and green innovations isn't just an equation for ecological salvation; it is likewise a guide for monetary development and headway. As the world wrestles with the outcomes of its past demonstrations, the capability of green advancements to change areas, make occupations, and drive economies is self-evident. Countries might clear a course to a more promising time to come by embracing maintainability as a driving variable for development. An opportunity to put resources into green innovations is presently on the grounds that the choices we make today will characterize the universe of tomorrow

Chapter 5

Agriculture and Food Systems: Feeding People and the Environment

5.1 Organic Farming: Putting Soil Wellbeing and Biodiversity First

Natural Cultivating: Putting Soil Wellbeing and Biodiversity First

In a time when the unrelenting walk of progress has regularly brought about accidental ecological impacts, natural cultivating fills in as an excruciating update that we can work with nature as opposed to against it. Natural cultivating is something other than a cultivating technique; an idea depends on veneration for the land, regard for environments, and a promise to people in the future. This philosophy is established on two indissoluble support points: soil wellbeing and biodiversity.

Think about going through a natural homestead, where the air is bursting at the seams with the murmur of bugs, bird singing, and leaf stirring. The soil underneath our feet is a no nonsense animal, overflowing with minute life frames that support the complex texture of presence above. The center of natural cultivating is live soil, a wellspring of force that requires food as opposed to substance control.

Natural makers trust in the deep rooted aphorism that solid soil produces sound harvests. They advance a fragile dance among microorganisms and plant roots by keeping away from engineered manures and pesticides. From mycorrhizal parasites to worms, these microorganisms assist with supplement cycling, bug control, and water maintenance. Subsequently, crops are more bountiful as well as more nutritious and delicious.

Be that as it may, the effects of natural cultivating go past the ranch. Soil, which is generally seen as a humble substrate, has developed into an integral asset in the battle against environmental change. Natural agrarian practices like cover editing and negligible culturing help the dirt's ability to hold carbon dioxide, which is basic in relieving the nursery impact. With the world wrestling with rising fossil fuel byproducts, the essential demonstration of cultivating soil wellbeing through natural strategies has taken on worldwide importance.

One more fundamental of natural cultivating is biodiversity, which recounts an account of strength and reliance. Monocultures prevail in customary farming, leaving crops helpless against sickness, bothers, and ecological vacillations. Natural horticulture, then again, makes a rich embroidery of life. Fields are decorated with an orchestra of plants, every one of which assumes an unmistakable part in protecting ecological balance. This natural concerto likewise incorporates hedgerows, local plants, and, surprisingly, useful "weeds."

A powerful urge to defend pollinators is at the core of natural cultivating's commitment to biodiversity. Honey bees, butterflies, and other pollinating bugs are unrecognized yet truly great individuals of our food framework, permitting many products of the soil to duplicate. Be that as it may, synthetic openness and territory misfortune present existential threats to these basic life forms. Natural homesteads that act as asylums offer them places of refuge, supporting a rebound of pollinator populaces that has expansive outcomes.

Besides, natural homesteads energize a social change that recognizes food as in excess of a product. Natural cultivating reasoning produces a feeling of association with the dirt as well as an enthusiasm for the critical effect of our choices. Ranchers, customers, and networks generally become stewards of a common environment connected by a common obligation to maintainability. Ranchers' business sectors, local area upheld farming (CSA) programs, and the developing interest for natural things on shop retires all mirror this attitude.

While the advantages of natural cultivating are enticing, obstructions remain. Natural strategies are often reprimanded for their more unfortunate yields and more noteworthy costs. These issues, notwithstanding, are progressively being handled with effective fixes. Through coordinated work on the board, accuracy in agribusiness, and the prudent utilization of cover crops, agroecological research is opening the potential for further developed efficiency.

At long last, natural cultivating welcomes us to look at our relationship with the Earth and the food we eat. It advises us that working with nature as opposed to against it is savvy and that by advancing soil wellbeing and focusing on biodiversity, we can prepare toward a more supportable future. Past the exquisite sights of succeeding homesteads and beautiful scenes, natural cultivating gives a model to ecological flexibility, a resurrection of conventional insight, and a reviving of our association with the regular world. As we stand at the junction of a quickly evolving globe, natural cultivating standards offer a way toward participation, concordance, and a future where overflow is as much about the essentialness of the land for what it's worth as the reap.

5.2 **Neighborhood Food Developments**: Overcoming any barrier Among Makers and Buyers

An exquisite unrest is fermenting in the heartlands of networks from one side of the planet to the other, one that goes past plain sustenance and reshapes the association among ranchers and customers. " Neighborhood Food Developments: Interfacing Makers and Buyers" plunges into this captivating gastronomic peculiarity, following its starting points, examining its effect on people and society, and featuring the manner in which it clears toward an additional feasible and interconnected world.

Making Change

Part 1: we follow the foundations of nearby food developments back to grassroots drives to restore failed to remember agrarian practices. These developments have created from humble starting points to become overall forces to be reckoned with, changing the manner in which we contemplate food, local area, and our relationship with the earth. We dig into the tales of the exploring ranchers, visionary campaigners, and enthusiastic clients who ignited the flash of progress, showing how their responsibility developed into a flourishing development.

A Comprehensive Encounter Past Taste

•**Part 2:** digs further into the tactile banquet that neighborhood food developments give. Beside the dazzling flavors, these developments give a vivid encounter that connects the faculties in general. We visit energetic rancher's business sectors, where clear tones, alluring aromas, and the murmur of local area impact. The section exhibits how neighborhood food is more than basically food — it's an excursion of disclosure, an ensemble of surfaces, and a lovely artful dance of flavors that makes a significant association between what's on the plate and the world's past.

The Harmonious Relationship

"<u>Neighborhood Food Developments:</u> Interfacing Makers and Purchasers" is truly about connections.

•**Part 3:** plunges into the cooperative relationship that creates between the people who develop the land and the individuals who partake in its prizes. Ranchers let us know firsthand about their encounters, from plowing the dirt to seeing their produce on supper tables. This part shows how these ties go past monetary trades, improving individuals' lives by means of shared encounters, shared convictions, and a solid obligation of common regard.

Local area Is Something beyond a Market

***Part 4:** the rancher's market is featured as an area to purchase food as well as a clamoring focal point of local area life. We catch the fervor, fellowship, and public soul that arises when individuals collect to help nearby makers through meetings and stories. We research the job of nearby food developments in encouraging conditions where cellphones are quiet and genuine discussions flourish — where the demonstration of buying a tomato is a chance to make connections that improve the social texture.

Gastronomic Resurrection

Neighborhood food developments have reformed the manner in which we eat out, from homestead to plate.

• **Part 5,** we visit the kitchens of ground breaking gourmet experts who are reshaping the culinary scene. We figure out how eateries are embracing the idea of nearby obtaining, making menus that move with the seasons. This section features the imagination of culinary experts who utilize neighborhood items to deliver exceptional dishes while likewise adding to an overall development for feasible eating.

Accommodating What's to come

•**part 6:** jumps into the tremendous effect of nearby food developments on the globe in a world adapting to natural worries. These developments give a guide to a stronger future by bringing down food miles, supporting reasonable practices, and shielding biodiversity. Each chomp we take turns into a decision in favor of a superior climate, a more equivalent food framework, and a general public where the choices we make during supper have extensive outcomes.

From Ranch to Fork: Buyer Strengthening

The clever wraps up with a message of strengthening for perusers. We feature the basic job that shoppers play in keeping up with nearby food developments

• **Part 7:** We give viable bits of knowledge that empower people to effectively partake in their food decisions, whether through buying at nearby business sectors, creating local area cultivates, or pushing for approaches that advance maintainable agribusiness. This section is a source of inspiration, empowering perusers to become change specialists, realizing that each chomp they move toward a more associated world.

"Nearby Food Developments: Interfacing Makers and Buyers" is in excess of a book; an assessment of a development gets to the core of nourishing our bodies, networks, and planet. Perusers will go on an excursion through pages celebrate flavors, make connections, and touch off the flash of progress, each significant piece in turn.

Credibility Flavors

•**Part 8:** we start on a culinary excursion that crosses limits and tastes. We research how nearby food developments honor social legacy through clear depictions and moving encounters. The part goes into the reclamation of notable harvests and recipes, welcoming light on how these developments give networks a stage to celebrate their culinary customs. This section presents a striking image of how neighborhood food developments keep social heritages alive, from legacy tomatoes that murmur recollections of ages past to recipes gave over through the ages.

The Pleasure of Patience

Nearby food developments move us to dial back and relish the second in our speedy climate. The joy of persistence that accompanies valuing the occasional rhythms of nearby items is investigated

•**Part 9.** We find how seeing strawberries age or pumpkins mature cultivates a profound appreciation for the regular world and its cycles. This section makes sense of how the expectation of a heavenly gather turns into a delicate update that a portion of life's most satisfying minutes create when we give them time through intriguing narrating.

Reaching out with Nature

Nature is essential to nearby food developments, and

•**Part 10** urges perusers to revive their relationship with the land. We travel with ranchers as they offer their points of view on feasible horticulture, biodiversity, and the need of natural stewardship. The part underscores the changing force of taking care of business, whether through local area gardens, school drives, or window-box spices. Nearby food developments, by reconnecting with the land, advise us that our prosperity is inseparably connected to the soundness of the environment.

•**Increasing: Deterrents and Wins**

While nearby food developments have had enormous achievement, they are not without troubles. To address the difficulties that these developments experience, going from contest with modern farming to dispersion challenges. Regardless of these difficulties, we track down instances of versatility and resourcefulness. We take a gander at how networks are consolidating to handle difficulties and adjust to changing scenes while staying consistent with their fundamental standards. As nearby food developments thrive and develop, this part shows the force of devotion and collaboration.

•Recipes for Change

In the last section, we complete the story by furnishing pursuers with a mother lode of recipes that catch the pith of nearby food developments. These recipes aren't just about making scrumptious food; they're likewise about injecting accounts of manageability, local area, and flavor into each nibble. Every recipe embodies the soul of "Nearby Food Developments: Associating Makers and Purchasers," empowering perusers to go on culinary journeys that interface the homestead and the table.

•Far in excess of the Plate

The epilogue provides perusers with a feeling of potential outcomes and trust. It plans ahead, envisioning a general public in which the thoughts of nearby food developments are significantly implanted. Networks are turning out to be more connected, ranchers are prospering, and the effect of our food decisions is causing great change on a worldwide scale. While the book is about food, it is eventually about the changing force of cooperating to make a more economical, scrumptious, and blissful world.

"Neighborhood Food Developments: Interfacing Makers and Purchasers" is in excess of a book; it's an encounter. It's an excursion through flavors, stories, and associations that help us to remember the astounding power that our food decisions have. It's a source of inspiration, encouraging perusers to enter an existence where each feast is an opportunity to celebrate local area, support the body, and develop a profound association with the earth. As they turn the pages, perusers leave on an enchanting experience that overcomes any issues between what we consume and the world we live in.

5.3 Food Security:

Guaranteeing Everybody Approaches Nutritious Food

Sustaining Trust: Propelling Food Security for a Sound World

A quiet battle seethes in this present reality where riches and destitution coincide, testing our shared perspective and raising doubt about the fundamental center of humanity. This battle is being pursued not on front lines, but rather in large number of individuals' kitchens, markets, and supper tables. The battle is for food security, which is an essential string that goes through the texture of worldwide prosperity. As we investigate the subject of "Food Security: Guaranteeing Admittance to Nutritious Nourishment for All," we will set out on an excursion to uncover the hardships, uncover the arrangements, and stir up change that will prepare for a general public wherein nobody heads to sleep hungry.

Exposing the Imperceptible Foe,

✓ **Section One**

The journey starts with an examination concerning the concealed fights that decide food security. The undetectable enemy of appetite and ailing health that torment people and networks hides underneath the cover of abundance. The main part dives into the stories of the people who face this trouble day to day, introducing a sensible image of the difficulty they experience. This section moves us to go up against the earnestness of the issue, from the country rancher attempting to develop to the point of taking care of her family to the metropolitan youngster whose fantasies are hampered by the aches of a vacant stomach.

✓ Section Two

dives into the secret of craving in the midst of abundance in a world flooded with food. We go into the perplexing trap of causes that add to this division, which goes from inconsistent circulation frameworks to monetary disparities that obstruct access. The part reveals the defects in our food frameworks, uncovering how, even during a time of specialized wonders, a large number of individuals stay caught in the grip of craving. Perusers are pushed to accommodate the truth of shortage in a universe of a lot through educational disclosures and charming tales.

The Strength Upheaval

✓ **Section Three**:

beams of trust show up among the difficulties. We tell the stories of people and networks who are driving the strength upheaval. This part features the advancement and devotion that fuel the battle against food weakness, from supportable cultivating rehearses that oppose environmental change to imaginative dissemination models that rise above metropolitan rustic partitions. Perusers are acquainted with areas of strength for the that each seed planted is a stage toward a yearning free future through the narratives of these change-producers.

✓ **section Four:**

Supporting Arrangements Through Approach

The focal point of Section Four is on the job of strategy in forming the direction of food security. We research the cooperation of public and worldwide strategies, analyzing how they could either reinforce or debilitate access snags. The section explores the labyrinth of strategy intricacies, from agrarian regulations to economic deals, showing how instructed and evenhanded approach making holds the way to eliminating the obstructions that stand among people and the food they expect to live.

✓ section five:

Cultivating People group Strengthening

As we research the groundbreaking effect of local area driven programs in Part Five, strengthening becomes the overwhelming focus. This section presents a clear image of how networks are taking responsibility for food security, from engaging ladies as change specialists to supporting farming cooperatives that help nearby versatility. Perusers experience the triumph of collective exertion in these tales and discover that genuine versatility starts when people cooperate to defy normal issues.

✓ section 6:

A Source of inspiration " An unquestionable necessity to do"

As perusers approach the finish of our examination in Section Six, they are welcomed with a source of inspiration — a call to be change engineers. This part exhibits that the battle for food security isn't just for policymakers and associations; it's a call for everybody to assume the job of progress producer. Each move made, from supporting neighborhood food drives to pushing for official changes, adds to the aggregate energy toward an existence where everybody approaches nutritious food.

✓ **section seven:**

The Cascading type of influence of Food Security

As we go into Part Seven, we'll take a gander at the monstrous wave influence that food security has. Food security feeds networks, organizations, and, surprisingly, the climate notwithstanding the body. We explore the perplexing connections between nutritious food access and improved wellbeing, instructive results, and financial result. The part accentuates the flowing impact of finishing the pattern of craving, exhibiting that when individuals are adequately taken care of, the advantages stretch out a long ways past their plates.

✓ Section Eight:

An Organization Ensemble, In Section Eight, we investigate the orchestra of unions that are critical in the street toward food security. This part investigates the force of cooperation, from legislatures and non-administrative associations to private area endeavors and grassroots endeavors. Perusers see the changing capability of alliances that rise above lines, areas, and belief systems through dazzling stories and contextual analyses, showing that the battle for food security requires the consonant assembly of different drives.

✓ section Nine :

Developing for a Craving Free Future

As we research the groundbreaking effect of innovation and imagination in impacting the eventual fate of food security, development arises as a main impetus in •Chapter Nine. This part shows how development is changing the game, from portable applications that associate ranchers to business sectors to vertical cultivating frameworks that change metropolitan agribusiness. Perusers are presented to the possibility that innovation and inventiveness can overcome any barrier among shortage and overflow through rousing models and game-evolving arrangement.

✓ section Ten:

Schooling and mindfulness arise as change specialists. We examine that it is so essential to bring issues to light about the intricacies of food security, scatter legends, and move instructed activity. The part makes sense of how instruction might empower individuals to pursue informed choices, advocate for change, and add to the production of a future where nutritious food is a widespread right. Perusers are reminded, by means of instances of training drives and grassroots missions, that information is a fundamental instrument in the excursion toward a more food-secure future.

✓ section Eleven:

Keeping up with Force, The penultimate section welcomes perusers to consider thoughts for keeping up with the energy of change. We examine the meaning of long haul responsibility, versatile strategies, and coordinating food security into bigger improvement objectives. This section gives an arrangement for proceeding with the direction toward a general public where admittance to nutritious food is definitely not a short lived accomplishment, yet a persevering through the real world, from evaluating progress to commending achievements.

At last, a Craving Free Heritage

As we close to the furthest limit of "Food Security: Guaranteeing Admittance to Nutritious Nourishment for All," the story finishes up with a strong message: We should leave a tradition of a craving free planet. We consider the narratives of boldness, the arrangements that have been ignited, and the cooperative endeavors that have been meshed into the texture of progress. Perusers are reminded all through this request that food security is a common vision that associates humankind in a typical mission. As we turn the last page, we are called to lead of backing, activity, and sympathy forward, guaranteeing that admittance to nutritious food turns into an enduring inheritance that we give to people in the future.

5:4 Plant-Based Diets:

Decreasing Ecological Effect and Advancing Wellbeing
In a world loaded with choices, the choice of what to put on our plates goes past just nourishment. It turns into an assertion — a declaration of values that can possibly influence our wellbeing, the climate, and the world we pass on to people in the future. " Plant-Based Diets: Limiting Ecological Effect and Advancing Wellbeing" dives into this groundbreaking excursion, uncovering the

powerful cooperative energy between our dietary decisions and the soundness of the world we call home.

On Our Plates: The Green Unrest

The excursion begins with an enchanting examination of the embodiment of plant-based eats less carbs.

Part One, we take a gander at the arising development that is reshaping our culinary world. The part uncovers the emotional change away from customary meat-driven feasts and toward a plenty of splendid plant-based choices. Perusers are presented to the universe of plant-based eating through enthralling narrating and enlightening bits of knowledge — a universe that is about something other than subbing creature items yet reevaluating the key way we ponder food Sustenance for the Body.

Part Two, the accentuation movements to the medical advantages that plant-based slims down give. We go over the experimentally demonstrated advantages of lower cholesterol levels, lower hazard of ongoing infections, and more prominent general prosperity. However, the story doesn't end there; this part investigates the consequences of private wellbeing. Each significant piece in turn, we explore how embracing plant-based food prompts an additional maintainable and sympathetic world.

The Manageability Ensemble

Part Three, we take a gander at the ensemble of manageability that plant-based consumes less calories give. The natural effect of meat-eating is obvious, from the enormous breadth of farming regions expected for animal creation to the amazing water and energy assets expected to keep up with it. This part directs perusers through the biological results of our food decisions, stressing how taking on plant-based diets can decisively bring down our carbon impression and make ready to a greener, stronger planet.

Released Culinary Inventiveness

Part Four uncovered a distinctive range of culinary development propelled by plant-based consumption of less calories. This part praises the variety and resourcefulness that flourish in the realm of plant-based eating, from delectable sans meat choices that rival their creature partners to the blast of plant-based cafés that are rethinking connoisseur. Perusers are encouraged to taste the captivating kinds of plant-based food through tales and enticing portrayals.

The Ethical Embroidered artwork.

Part Five, we examine the ethical part of our eating decisions, where morals and empathy cross. We dig into the existence of creatures and the moral scrapes that modern creature creation presents. This section urges perusers to investigate how their decisions interface with their qualities and desires for a more empathetic world by illuminating creature mercilessness and the ecological harm it causes.

Progress Controlled by Plants

We honor the surprising accounts of people, networks, and associations propelling plant-fueled change. This section shows how plant-based eats less are not only a pattern, but rather an extraordinary development, from grassroots drives to corporate supportability vows. Perusers will see the joined impact of differed endeavors to make a better, more manageable, and kinder future through these stories.

Disposing of Fantasies and Misguided judgments

The normal false impressions and misguided judgments about plant-based eats less carbs. We address incessant worries about protein consumption, nutrient shortages, and flavor. We scatter these fantasies with proof based clarifications and master sees, revealing insight into how very much arranged plant-based diets might supply each of the necessary components for a solid way of life. This part gives perusers the information they need to serenely explore discussions and pursue informed choices.

Creating Cognizant Utilization

Part six : urges perusers to fabricate a more mindful way to deal with food utilization. We dive into the idea of careful eating, encouraging perusers to enjoy each chomp, associate with the wellsprings of their food, and contemplate the bigger ramifications of their choices. This part advises us that each feast is a chance to support our bodies as well as our souls, from savoring the surfaces and flavors to saying thanks to crafted by the people who develop the food.

Plant-Based Diets for Every Occasion

Part seven: , we feature plant-based diets' versatility and the way in which they can be altered to various lives and interests. A plant-based diet gives versatile options whether you're a competitor searching for max operation or a bustling proficient exploring furious days. We see how plant-based diets might be remembered for social traditions, family suppers, and other culinary customs, exhibiting that doing the switch isn't just imaginable yet in addition to improving.

The Street Ahead

As the story approaches its peak in Part seven, perusers are encouraged to stop and ponder their excursion. The section gives a guide to moving the energy of change along. We take a gander at pragmatic strategies for taking the action to a plant-based diet, for example, dinner arranging, shopping for food ideas, and dishes that feature the range of choices accessible.

Each Chomp In turn Change Supporting

As we close to the furthest limit of our excursion, the message turns out to be clear: the excursion of embracing plant-based counts calories is a convoluted movement that impacts all components of our lives. Plant-based diet welcomes us to become dependable stewards of our bodies and the Earth, from reestablishing our wellbeing to patching the climate and maintaining our moral standards.

It's not just about what's on our plates; it's tied in with leaving a tradition of harmony, empathy, and respect for the fragile texture of life.

"Plant-Based Diets: Limiting Natural Effect and Advancing Wellbeing" closes with a source of inspiration that reverberates a long ways past the book's words. The end builds up that our food decisions are something beyond private inclinations; they are woven into the more extensive embroidery of the globe. By taking on plant-based slims down, we help to make a future in which wellbeing and maintainability coincide, empathy and care describe our cooperations with creatures and the climate, and the heritage we leave behind is one of positive change.

At the point when perusers finish this investigation's part, they aren't simply completing a book; they are leaving on a changing experience. They are entering a world in which each dinner is a chance for strengthening, for supporting their bodies as well as the climate around them. Plant-Based Diets: Limiting Natural Effect and Advancing Wellbeing" is an encouragement to be the change we need to see — a call to set out on an invigorating way loaded up with imperatives.

Chapter 6

Responsible Consumption and Production: Changing Habits

6.1 Mindful Consumption: Focusing on Better standards without ever compromising

It's not difficult to fall into the snare of careless realism in a world that ceaselessly besieges us with ads, deals, and the allure of the most recent patterns. We are apparently gathering merchandise at a disturbing rate, frequently failing to focus on the fundamental worth and motivation behind our buys. In any case, a huge shift is occurring — one that advances care in consuming and the idea of value above amount. Welcome to the time of careful commercialization, in which each buy turns into a cognizant choice and the objective of long haul satisfaction bests short lived guilty pleasure.

Cognizant Utilization Is Arousing

Careful commercialization is certainly not a brief pattern; rather, a response to the mind-boggling overabundances have portrayed the purchaser climate for quite a long time. We're beginning to acknowledge as purchasers that more doesn't rise to better 100% of the time. This enlivening is spurred by a longing for more profound binds with our assets and a superior comprehension of the impact of our choices on both our lives and the earth.

In a world doing combating with ecological issues going from environmental change to asset consumption, we are finding that our buying propensities are basic. Each item we purchase has a history: the assets expected to make it, the work engaged with its assembling, and the ecological effect it abandons. Careful industrialism urges us to go underneath the surface and inspect these stories prior to settling on choices that influence our prosperity as well as the prosperity of our reality.

The Quest for Life span, Envision a world in which, rather than gathering a plenty of modest and dispensable items, we put resources into less, better belongings that are dependable. This is the substance of insightful commercialization — choosing higher expectations without compromise to ensure that what we buy satisfies its motivation effectively as well as endures. At the point when we pick quality, we limit squander and the requirement for incessant substitutions, diminishing our commitment to the expendable culture that has turned into the norm.

Quality things are something other than objects; they have natural worth that goes past their financial worth. A very much created calfskin sack, for instance, conveys the producer's expertise and capability, and each wear addresses an unmistakable excursion. A hand-cut feasting table turns into a point of convergence for family social events and celebrations. Careful commercialization urges us to foster associations with our assets, perceiving their significance in our lives and developing a feeling of satisfaction that rises above realism.

How frequently have we gotten ourselves quickly adding items to our internet shopping crates or carelessly swiping our installment cards? Careful commercialization urges us to stop, consider, and assess our inspirations. Is the buy persuaded by a genuine need, or is it the consequence of a keen promoting effort? By practicing care, we can recognize momentary cravings and certified necessities, permitting us to make buys that are predictable with our qualities and desires.

Moreover, smart utilization urges us to assess the more extensive results of our choices. Is the organization we're backing moral and long haul? Do they put a superior on fair work principles and naturally feasible assembling processes? These inquiries empower us to mindfully utilize our buying power, empowering firms to embrace more moral practices and add to everyone's benefit.

As the cognizant commercialization development gains forward momentum, it interfaces with another changing way of life — moderation. Moderation isn't tied in with living with as little as could be expected; rather, it's about deliberately adjusting one's life to what matters. We normally float toward a moderate disposition when we center around better standards without ever compromising, where products are intentionally picked in view of their worth and importance to our lives.

The opportunity that accompanies straightforwardness is a welcome help from the obligations of inordinate possession. We set aside physical and profound space for encounters, connections, and self-awareness by shedding the heaviness of superfluous merchandise. This freedom stretches out past substantial belongings; it influences our monetary choices, using time productively, and in general prosperity. We assume the job of life modelers, making a story that advances satisfaction over gathering.

Embracing Careful Industrialism, The course to careful commercialization is a steady development that needs care, mindfulness, and an eagerness to change. Everything begins with a change in context — a purposeful choice to esteem quality above amount in all pieces of our life. We not just further develop our prosperity by going with insightful choices that address our convictions, yet we likewise add to an additional supportable and simply world.

In a culture that much of the time partners satisfaction with material things, cognizant commercialization compels us to rethink our meaning of happiness. It empowers us to track down get a kick out of significant occasions, enduring associations, and the information that our choices have an effect past our nearby circles. As we progress forward with this excursion, remember that each buy is a chance to have a constructive outcome, support capable practices, and impact a future in light of the standards of careful utilization.

The Careful Commercialization Expanding influence, The effect of careful industrialism arrives at well past our own lives; it causes a far reaching influence that resounds through whole organizations and networks. At the point when we stress better standards no matter what, we send major areas of strength for a to producers and retailers that we anticipate dependable things. This change popular may incite associations to rethink their assembling works on, moving away from arranged oldness and toward a promise to solidness.

Organizations that answer this requirement for life span every now and again utilize conventional workmanship and feasible materials, advancing the rebuilding of distinctive abilities and nearby economies. This restoration of value driven creation can possibly renew regions that have been obscured by large scale manufacturing and reevaluating. Generally, careful industrialism catalyzes positive change, recovering enterprises and shielding social legacy.

Careful Utilization, Feasible Future, At its heart, careful commercialization is an interest in a more maintainable future. It perceives that the choices we make today impact the planet we depart for people in the future. We can bring down the weight on our planet's assets and diminish the ecological effect of our ways of life by moving our disposition from expendable to sturdy, from amount to quality. Individual decisions can have a groundbreaking impact, prompting a cultural development toward dependable commercialization and a better planet.

Consider a future wherein landfills are done spilling over with squandered items, seas are liberated from plastic contamination, and assets are used carefully and productively. This vision of a supportable future is feasible, and it begins with our cognizant decisions as cognizant purchasers. Each time we purchase a very much made, enduring thing, we help to make a more economical pattern of creation, utilization, and removal.

Developing Careful Commercialization, Taking on careful commercialization requires a cognizant work on in our propensities and perspectives. Here are a few substantial procedures to developing this changing outlook:

Think about Your Qualities: Carve out opportunity to consider what means quite a bit to you. Ponder the natural, cultural, and individual ramifications of your buys. Permit these qualities to direct your navigation.

Careful Independent direction: Stop and consider prior to making a buy. Consider whether the thing resounds with your qualities and gives a genuine motivation in your life.

Explore and Ask: Learn all that you can about the items you're pondering. Inspect the organizations' practices, morals, and ecological exercises. Help organizations that esteem capable creation.

Focus on Life span: Select items that are notable for their quality and sturdiness. While they might be more costly front and center, they often offer unrivaled benefit over the long haul.

Diminish Indiscreet Purchasing: Oppose the motivation to make rash buys. Permit yourself sufficient opportunity to consider whether the thing is really important and finds a place with your careful commercialization points.

Clean up With Goal: As you shift to a more cognizant way of life, clean up your space with expectation. Give or reuse merchandise you never again need, and let go of the aggregating attitude.

Support Nearby and High quality: Search for things carefully assembled by nearby craftsmans or little organizations. These things regularly have a novel story to tell and add to feasible economies.

Teach and Supporter: Illuminate your social associations about careful industrialism. Empower open exchange on the outcomes of purchaser decisions and promoter for capable way of behaving.

A Tradition of Cognizant Decisions, Careful utilization is more than basically a craze; it is an inheritance we leave for people in the future. A tradition of conscious decisions exhibits our commitment to a more significant, feasible, and satisfying lifestyle. We rebuild our relationship with material merchandise by picking higher standards no matter what, reconnecting with the genuine embodiment of worth, and embrace a future in which our activities address our deepest feelings.

As we leave on this street of cognizant shopping, let us recollect that we can influence change, both in our own lives and in the bigger world. Each purposeful buy is a chance to interface our ways of behaving with our objectives, to focus on morals over comfort, and to commend the craft of careful living. Each choice we make shapes our stories as well as the more extensive story of our general public — an account that repeats the insight of picking higher standards without compromise.

6.2 - Utilize Plastic Decrease:

From Attention to Activity - Lessening Single-Use Plastics: From Attention to Activity, Envision a world with clean sea shores extending to the extent that the eye can reach, seas spilling over with rich marine life, and air liberated from the devastating handle of plastic contamination. This isn't some distant dream; it is a potential future. The way from attention to activity in dispensing with single-use plastics is an endeavor that humankind should embrace for our planet's wellbeing, prosperity, and the drawn out endurance of a thriving environment.

The Enlivening: A Plastic Plague, Mindfulness is the most vital move toward change, and the world has as of late awoken to the horrendous truth of single-use plastics. These dispensable wonders, once hailed for their straightforwardness, have now turned into an ecological calamity, dirtying our seas, killing marine life, and entering even the most remote corners of the Earth. Plastic straws, sacks, containers, and bundling have become connected with an expendable culture that has had a huge ecological effect.

Narratives, news reports, and virtual entertainment have carried the plastic fiasco to the front of our consideration, dazzling us with pictures of seabirds caught in plastic rings and turtles gagging on plastic sacks. This expanded information has ignited a collective sensation of obligation, convincing us to stand up to the disrupting reality that our comfort comes at a frightening expense. A reminder calls for activity.

From Lack of concern to Backing: Lighting Change with Energy and Reason Lack of concern can emerge as a quiet enemy in a world frequently wrecked by the bedlam of day to day presence, creating a shaded area over the chance of positive change. However, inside every one of us is a flash of backing, a strong power fit for changing lack of care right into it and detachment into impact. The excursion from disregard to promotion is an account of change wherein customary individuals utilize their interests to impact extraordinary change.

Unresponsiveness, which is normally the consequence of defenselessness or thwarted expectation, is a barrier to headway. It proposes that our exercises are unimportant and that the difficulties we face are too enormous to even think about surviving. Nonetheless, history is loaded up with instances of the individuals who separated this obstruction — individuals who changed their feelings into fuel and their questions into drive. These activists, propelled by a powerful urge to have an effect, show that a solitary voice, a solitary idea, or a solitary activity might redirect our planet.

The excursion starts with an enlivening — a second when lack of concern's murmurs are overwhelmed by heart's yelling. A comprehension quietness isn't a choice in a world loaded with difficulties. Backing emerges as the counteractant to lack of care, getting a fire going inside us that won't go out. Promotion, whether for natural maintainability, civil rights, or common freedoms, is a persistent commitment to defending what is correct, intensifying the voices of the oppressed, and testing the current quo.

From attention to activity, backing overcomes any issues among smugness and change. Concerns become missions, and shock becomes projects. It unites individuals who won't be uninvolved spectators and accept that a superior future is a common commitment, not a far off great. Support joins networks, manufactures alliances, and fashions developments fit for reshaping approaches, businesses, and societies.

Support, then again, isn't invited all the time. It is met with doubt, resistance, and affliction. Regardless, the promoters' energy is a strong power that can't be held back. Backing, similar to a guide in the evening, illuminates bad form, blending hearts, rousing discussions, and provoking activity. It considers those in power responsible, advising them that they are liable to individuals they serve.

Backing is not generally restricted to huge motions or conspicuous stages in the present connected society. It blossoms with standard demonstrations of courage, whether they be revolting against bias, helping psychological wellness mindfulness, or making discussion on urgent subjects. Each demonstration of campaigning sends swells through society, molding mentalities, discernments, and strategies.

The way from lack of concern to support is pretty much as interesting as individuals who take it. Advocates come in many shapes and sizes, from grassroots coordinators to worldwide powerhouses, from neighborhood activities to global developments. What unites them is their hesitance to acknowledge the current quo and their undaunted conviction that they can roll out an improvement, regardless of how little or huge.

Age, occupation, or history are not boundaries to promotion. It's a mantle that anybody can get, a light that can be passed down from one age to another. The tradition of progress stays as the present activists rouse the upcoming pioneers, a recognition for the getting through strength of the human soul.

At long last, the excursion from lack of care to activism embodies human potential. It exhibits our capacity to battle aloofness with enthusiasm and misery earnestly. An update even the littlest activities can have an enormous effect, that a solitary voice can be heard across seas, and that our choices today can change the universe of tomorrow.

Thus, let us answer the source of inspiration. Give us turn our irritations access to fuel, and our compassion right into it. In a world longing for change, we can make our voices heard, to battle for equity, and to leave an enduring heritage. The excursion from lack of care to promotion is a festival of the dauntless soul that propels us to impact the world each significant stage in turn.

The Force of Buyer Decision, The force of shopper decision is at the core of this transformation. We vote in favor of manageability each time we pick a reusable water bottle for an expendable one or use fabric packs at the supermarket. The repercussions of these choices are huge. At the point when a great many individuals go with cognizant choices, businesses are compelled to adjust, encouraging an interest for eco-accommodating items and bundling.

Moreover, customer pressure is making brands reexamine their assembling systems. Organizations that previously cheered in single-utilize plastic bundling are progressively putting resources into examination and development to find choices that keep up with comfort while bringing down their ecological impact. We influence the economy as well as the principal texture of our way of life, where the expendable mentality is giving way to a devotion to stewardship.

Little Advances, Enormous Effect, The way from information to activity doesn't necessarily in every case incorporate extraordinary motions; some of the time, little advances have the most significant effect. Think about saying "no" to a plastic straw or conveying a reusable espresso mug. These apparently unimportant endeavors structure the underpinning of a worldview change. They mean a shift away from neglectful utilization and prepare for smart choices with broad outcomes.

These minor estimates go past individual activities. Schools are showing kids the plastic fiasco, in this manner empowering another age of naturally scrupulous residents. Local area tidy up endeavors are gathering volunteers to free stops, sea shores, and waterways of plastic litter. People who partake in these occasions become change envoys, accentuating the message that cooperative exertion can address even the most considerable issues.

The Street Ahead: Difficulties and Wins, As we move from attention to activity, we face provokes that put our will under a magnifying glass. The simplicity of single-use plastics, as well as the foundation that empowers their creation and conveyance, stay imposing rivals. To beat these obstructions, new thoughts, creative innovations, and a unified overall front are required.

In spite of the challenges, a few triumphs enlighten the way forward. The development of "zero waste" social orders, in which individuals attempt to diminish their garbage yield, exhibits human imaginativeness and assurance. Inventive undertakings, for example, plastic-eating microorganisms and sea cleaning devices, exhibit our capacity to plan arrangements that address the plastic test at its root.

A Source of inspiration: Our Aggregate obligation, to lessen single-use plastics isn't just the obligation of states, organizations, or natural associations — it's a source of inspiration for everybody. We are partners in this battle, and our choices, of all shapes and sizes, decide the course of our reality. An obligation cuts across lines, nations, and ages, joining us through a common devotion to our planet's prosperity.

The Force of Now, This second is vital in the radiant embroidered artwork of mankind's set of experiences. The excursion from attention to activity is a basic stage, deciding the heritage we leave for people in the future. The choice is clear: We can either draw out the plastic calamity by sitting idle or add to its goal by settling on purposeful decisions.

At long last, the excursion from attention to activity shows human organization. It fills in as an update that our activities can possibly change our reality. We declare our obligation to a cleaner, better, and more lively Earth by diminishing single-use plastics. It's a journey of confidence, development, and aggregate change that people in the future will recollect.

6.3 E-Squander The board:

Gadgets Reusing and Reuse, E-trash The board: Reusing and Reusing Gadgets, In the computerized age, innovation progresses at a staggering rate, coming about in a steadily developing pile of electronic trash, or e-squander. As our lives become more snared with gadgets and thingamabobs, great e-squander the board is a higher priority than any time in recent memory. Reusing and reusing hardware isn't only really great for the climate; it is likewise a stage toward a more feasible future.

Step inside any home or business, and you'll more than likely track down a large number of electronic gadgets. Our lives are firmly laced with innovation, from PDAs and PCs to TVs and family gear. While these gadgets are helpful, they likewise add to a serious worldwide issue: e-waste. E-squander is comprised of deserted electronic gear that breeze up in landfills or incinerators, where hurtful mixtures are delivered into the climate. Electronic gear is becoming outdated at a disturbing rate because of the fast speed of specialized upgrades, expanding the e-squander issue.

The Natural Effect, The ecological results of improper e-garbage removal are serious. Lead, mercury, cadmium, and brominated fire retardants are among the poisonous mixtures tracked down in hardware. These contaminations debase environments and posture serious wellbeing worries to the two people and natural life when they saturate soil and water. Besides, the development of electrical gear requires a significant measure of energy and materials. We can decrease the interest for unrefined components and energy-concentrated assembling processes by reusing and reusing contraptions.

Reusing Hardware: Finding Stowed away Fortunes Reusing hardware involves fastidiously dismantling contraptions to gather important parts like valuable metals, copper, and plastics. These materials can be reused and used to make new devices. Reusing hardware is basically equivalent to mining a rich crease of minerals from out of date contraptions. We can recuperate roughly 24 kilograms of gold, 250 kilograms of silver, and a few kilograms of copper by reusing a lot of phones, for instance. This recoveries significant assets as well as diminishes the requirement for earth risky mining exercises.

The Craft of Reuse: Broadening Lifecycles, While reusing is a significant piece of e-squander the board, the specialty of reuse becomes the overwhelming focus. Reusing hardware expands their lifecycles, bringing down the all out need for new gear. Giving practical hardware to schools, local area associations, or restoring projects can assist with connecting the advanced separation and furnish ruined bunches with admittance to innovation. Moreover, renovated hardware could find new homes in districts where pristine things might be far off monetarily. This works on the climate as well as friendly value.

Inventive E-Squander The board Approaches, The battle against e-squander is certainly not an uneven one. Across the globe, one of a kind arrangements are being created by creators. Innovation is having a basic influence in tending to the e-squander challenge, from secluded telephone plans that make support and overhauling a breeze to eco-accommodating materials that reduce the natural effect of assembling. States and associations are additionally reaching out, creating e-squander assortment and reusing programs, laying out drop-off areas, and implementing stricter e-garbage removal regulation.

A Source of inspiration: Our Job in E-Squander The board, People assume a significant part in e-squander the board. Everything starts with cognizance. Understanding the outcomes of unseemly removal can rouse us to use sound judgment. At the point when it comes time to express farewell to our old gadgets, reusing or giving them ought to be the standard as opposed to the exemption. Numerous hardware stores and producers have reclaim projects to guarantee that your old gadgets are appropriately discarded.

The Commitment of a Feasible Future, Envision an existence where innovation's tenacious walk doesn't come to the detriment of our earth. Each cell phone overhaul, PC switch, or television substitution assists with making the world a greener, cleaner place. E-squander the board is about something other than waste administration; it is tied in with clearing the street for a more maintainable future. We are not just unloading the old by reusing and reusing contraptions; we are introducing another period of ecological cognizance and moral advancement.

E-squander the board is a basic string in the wide woven artwork of natural worries. As our lives become more computerized, the need of appropriate gadgets removal and utilization can't be worried. Reusing and reusing hardware is something beyond lessening garbage; it is an explanation that we value the strength of our planet however much we esteem mechanical wonders. In this way, the following time you update your gadget, remember that your choices can assist with characterizing a future wherein innovation and manageability coincide together.

Embracing the E-Squander Upheaval, envision a reality where obsolete cell phones are changed over into clinical contraptions, where disposed of PCs track down new life in instructive organizations, and where the previous hardware drives the upcoming disclosures. This is the captivating vision of the e-squander upset, a developing development in which people, partnerships, and legislatures team up to change the story encompassing electronic junk.

changing over junk into Fortune, Albeit the idea of changing over garbage into treasure isn't new, it has taken on a totally different significance with regards to e-squander. Our electronic gadgets' parts are stacked with conceivable outcomes. Disposed of motherboards can be dug for their muddled hardware, which can be used to deliver new electrical parts if appropriately recuperated and sanitized. This diminishes the necessity for new materials while likewise preserving energy and cutting ozone depleting substance discharges.

Moreover, e-squander reusing isn't restricted to metal and plastic reusing. It's additionally about protecting uncommon earth components, which are utilized for the development of all that from wind turbines to electric vehicles. As these components develop all the more scant, reusing them from obsolete gadgets becomes financially reasonable as well as basic for speeding up the spotless energy progress.

The Human Association, E-squander the board is an account of human association and strengthening, not just circuits and metals. Consider the professional who purposefully revamps a dismissed tablet, giving it a renewed outlook in the possession of an energetic student. Consider the towns that get by on e-squander gathering and reusing. Casual e-squander reusing networks have jumped up in unfortunate countries, extending employment opportunities for the people who gather, destroy, and sort electrical devices.

In any case, the casual idea of these tasks habitually includes some significant pitfalls: laborers are much of the time presented to poisonous materials without a trace of fundamental security techniques. We can safeguard both the climate and individuals who add to e-squander reusing by formalizing and controlling these cycles.

Planning for a Roundabout Economy, A change in plan mentality is expected to really address the e-squander issue at its base. Enter the roundabout economy idea. A round economy, instead of the exemplary direct worldview of "take, make, arrange," imagines a persistent circle of asset use and recuperation. This involves planning gadgets in view of repairability and upgradeability with the goal that gadgets might be handily fixed and parts supplanted on a case by case basis.

Consider a cell phone with particular parts: a harmed screen could be supplanted, and the battery could be improved without supplanting the whole gadget. Such plans increment the life expectancy of hardware as well as lessen asset strain and the heap of e-squander.

Teaching for Change, As the e-squander the board development gets pace, schooling is turning out to be progressively significant in advancing change. Bringing issues to light about the repercussions of ill-advised removal can propel individuals to act. Schools and colleges can integrate e-waste and supportability addresses into their educational programs, sustaining an age that knows about the natural outcomes of their choices.

Besides, producers ought to adopt a proactive strategy by offering data about their products' lifecycles, remembering particulars for reusing decisions and repairability. Very much educated customers have enabled shoppers, ready to pursue cognizant choices that wave across supply chains and effect industry conduct.

The Following stages, The way to successful e-squander the board is a long distance race, not a run. It requires the concurrence of innovation advancement, administrative change, purchaser mindfulness, and modern responsibility. While obstructions stay, the advancement acquired is verifiably empowering.

The scene of e-squander the board is changing, from metropolitan mining to eco-accommodating hardware make. The center of this development, be that as it may, is in the everyday choices we make - the choice to reuse as opposed to squander, to fix as opposed to supplant, and to advocate for reasonable practices. We can change the fate of innovation by embracing the e-squander transformation, making a story in which improvement doesn't come to the detriment of our reality.

E-squander the board is an orchestrating note in the orchestra of natural protection, advising us that the capacity to impact change is immovably in our grasp. As the globe advances toward an undeniably computerized future, let us ensure that the reverberations of our electronic strides leave a tradition of dependable utilization, faithful removal, and a planet saved for people in the future.

6.4 Maker Responsibility: Considering Organizations Responsible for Squander

Broadened Maker Obligation: Considering Organizations Responsible for Squander, In this present reality where utilization rules, a mystery issue - a waste emergency - has been building. Heaps of rubbish, going from plastic bundling to innovative gadgets, are gagging landfills and debasing the seas. As the need to address the waste problem rises, an influential idea known as Expanded Maker Obligation (EPR) has developed. EPR turns the tables on squander the board, empowering organizations to take responsibility for merchandise's entire lifecycle, from support to grave. This sensational shift is more than essentially an answer; a progressive methodology considers organizations responsible for their items and enables a future where waste isn't the cost of development.

Another Responsibility Worldview, Envision a world where the moment you purchase a thing, its destiny is fixed. It is as yet attached to the producer, ensuring that it doesn't turn into an ecological weight. That is the embodiment of Expanded Maker Obligation, a worldview change expecting organizations to effectively deal with their items all through their life cycles. This involves making things that are easy to dismantle, reuse, and reuse. It involves verifying that the materials used are alright for the climate and that finish of-life removal is painstakingly made due.

The limits between creation, utilization, and waste are obscured by EPR. It's a trying deviation from the proven "take-make-arrange" technique. Organizations are presently responsible for the entire life pattern of their items, from the industrial facility floor to reusing or safe removal.

Shutting the Waste Circle, The brightness of EPR rests in its capacity to close the waste circle. Customers have generally borne the brunt of rubbish the executives, arranging, reusing, and discarding items created by organizations. Be that as it may, with EPR, the weight is gotten back to where it should be: the actual organizations. This urges them to contemplate the natural effect of their items all along, bringing about development in eco-accommodating plans and materials.

Take, for instance, plastic bundling. Organizations subject to EPR should investigate choices, for example, biodegradable bundling or empowering reusable compartments as opposed to producing single-go through plastics that end choking the world. This isn't simply a restorative shift; an unrest powers organizations to think long haul and embrace supportability.

A Mutually beneficial Arrangement for the Climate and the Economy, Some might contend that making firms obligated for waste will block development and weight them with extra consumptions. The fact, be that as it may, is an incredible opposite. EPR benefits both the climate and the economy.

First of all, associations that embrace EPR are regularly at the forefront of development. They open up a universe of additional opportunities by imagining things that are easy to dismantle and reuse. Development flourishes in reusing innovation, reusing drives, and shut circle frameworks. Organizations that esteem EPR are regularly commended for their ecological administration, yet in addition for their imagination.

Second, the financial benefits of a cleaner climate are limitless. The expenses of tidying up after inefficient ways of behaving, managing natural mischief, and overseeing junk mountains far offset any early interests in manageable strategies. Besides, as buyers request all the more harmless to the ecosystem items, organizations that share these qualities gain an upper hand, drawing in cognizant purchasers who need to pursue capable decisions.

The Obligation Far reaching influence, EPR modifies something other than items; it changes attitudes. Organizations that embrace this responsibility idea lay out a huge point of reference for areas and clients the same. A culture shift toward careful buying and environmentally mindful choices is the aggregate gradually expanding influence.

Consider the gadgets business, which is known for its high item turnover and resulting e-squander. EPR urges gadgets makers to make items that are effectively repairable, upgradeable, and recyclable. This decreases e-squander as well as adjusts purchaser view of gadgets.

Rather than survey gadgets as expendable wares, clients are starting to see the value in the craftsmanship and strength of enduring items.

Regulations and the Authoritative Scene, The street to full-scale EPR reception isn't without troubles. Regulation is basic in cherishing EPR ideas in regulation. Numerous areas have found a way significant ways to institute regulation expecting firms to acknowledge liability regarding their items. These regulations even the odds, guaranteeing that moral organizations don't confront uncalled for contest from the individuals who worth momentary benefit over long haul supportability.

Regulation likewise guarantees that EPR is a legitimately restricting necessity instead of a deliberate commitment. It coordinates ecological still, small voice into business technique, introducing a future where capable ways of behaving are as of now not an upper hand yet the business standard.

Broadened Maker Obligation is a vow for a superior tomorrow in a time of gigantic utilization. It is a vow to shield our reality from the invasion of trash, as well as a guarantee to people in the future that progress doesn't need to come to the detriment of their legacy. EPR communicates the idea that organizations are something beyond producers; they are likewise stewards of the World's assets.

Organizations that take on EPR expect a future where waste is presently not an approaching misfortune yet a sensible undertaking. It is a world wherein development is compelled by ecological stewardship, and the things we make presently don't torment us as waste tomorrow. The world is awakening to the significance of this change, and as the flares of progress are fanned, we draw nearer to a future where responsibility isn't a choice; it's the main way forward.

Enabling a Round Economy Through Expanded Maker Obligation, Broadened Maker Obligation (EPR) is a string that sews together a few components of maintainability in the tremendous embroidery of waste administration. An innovation empowers us to move from a direct to a roundabout economy - an economy wherein assets are safeguarded, items are made for life span, and waste is decreased. EPR is in excess of an idea; it is a development at the convergence of natural mindfulness and business responsibility.

Planning for Dismantling: A Change in outlook, At the focal point of the EPR idea is the guideline of "plan for dismantling." Generally, things were intended to be utilitarian during their lifecycle and disposed of at the end. This content is flipped by EPR, which urges planners to make things that can be just dismantled and their parts reused or reused.

Consider a world wherein cell phones are not fixed closed, and where each nut and screw in a household item plays a part even after its unique shape is not generally needed. This shift toward measured and repairable plan broadens item life as well as limits the interest for natural substances and energy utilization. Organizations that adopt this strategy focus on maintainability instead of simply making things.

EPR Hugely affects Client Conduct, EPR doesn't simply consider firms responsible; it likewise hugely affects client conduct. Purchasers are more disposed to assess the total lifetime of what they purchase when things are made with the assumption for moral removal. This change in context begins a conversation on the worth of possession versus superfluity.

Consider the design business, which is known for its quick moving cycles that produce recent fads while disposing of old ones. EPR-motivated cycles might bring about the advancement of attire that might be effortlessly dismantled for reusing or reusing. This not just lessens how much waste in landfills yet in addition changes how shoppers see clothing. Rather than pursuing transient style, they put resources into pieces with long haul worth, prompting a more maintainable design climate.

EPR: From Hypothesis to The real world, Rejuvenating the idea of EPR requires a diverse methodology. State run administrations assume a significant part in ordering EPR rehearses across businesses by passing and upholding regulation. These principles force organizations to assimilate the full expense of their items, considering the natural effect.

Joint efforts between state run administrations, enterprises, and non-legislative associations can likewise cultivate advancement and best practices reception. To lay out a comprehensive arrangement, partners from makers to squander the board offices should be engaged with this conversation.

Worldwide EPR Points of view, Nations from one side of the planet to the other are driving the way in EPR execution, each with their point of view. In Europe, for instance, the Waste Electrical and Electronic Hardware

(WEEE) order forces on gadgets makers the obligation of dealing with the removal and reusing of their items. The Follow up on Advancing the Reusing of Little Waste Electrical and Electronic Gear in Japan urges makers to gather and reuse their items.

China, a significant supporter of worldwide waste, has as of late started an excursion to execute EPR standards by restricting imports of certain recyclables and advancing homegrown reusing drives. These models show the different manners by which EPR can be custom fitted to the financial and natural settings of different areas.

The Street Ahead, The street ahead isn't without its concerns. EPR execution requires a change in perspective in plans of action, supply chains, and customer discernments. Organizations should put resources into Research and development to make round things. Shoppers should move from an expendable culture to one of stewardship. States should sanction clear guidelines and structures that consider organizations responsible for the items they produce.

In any case, amazing open doors exist inside these difficulties. The valuable chance to start advancement, make long haul occupations, and safeguard our planet is gigantic. EPR is a basic player in the progress to a greener future, one where each item's lifecycle is carefully coordinated and squander is seen as an asset ready to be recovered.

A Vow for People in the future, Expanded Maker Obligation is a reverberating harmony in the terrific symphony of natural stewardship that resounds with the upsides of maintainability and corporate morals. It's a commitment made to current ages, yet in addition to people in the future, that the mix-ups of the past won't characterize our heritage.

EPR sparkles as an encouraging sign as the world wrestles with the outcomes of wild industrialism and unrestrained waste. It's an update that each step includes in the sensitive dance among industry and nature. An energizing cry cuts across businesses and lines, encouraging us to fundamentally impact the manner in which we make, consume, and discard items.

Chapter: 7

Global Collaboration: Tending to Worldwide Difficulties

7.1 The Assembled Countries Reasonable Advancement Objectives: A Guide for Worldwide Activity

In a world plagued by difficulties going from environmental change to neediness, the Unified Countries' Maintainable Improvement Objectives (SDGs) act as an undaunted light of trust and a source of inspiration. The SDGs, imagined as a thorough structure for settling the world's most earnest emergencies by 2030, address mankind's common objectives for a more libertarian, just, and manageable future. They are something beyond an assortment of goals; they address a shared promise to abandon nobody and to clear a way toward a future where thriving isn't an honor however an inheritance.

Groundbreaking Vision

The 17 Sustainable Development Goals, sometimes known as the Global Goals, represent a groundbreaking vision that rises above boundaries and belief systems. Every objective is a revitalizing point, an essential achievement focused on a particular part of world development. The SDGs incorporate many human necessities and desires, from destroying destitution and yearning to giving satisfactory training and orientation balance. They advocate for a finish to natural weakening, the improvement of serene and just networks, and the development of organizations to catalyze progress. In doing so, the objectives recognize the interconnection of our challenges and the need for cooperative arrangements.

The Impact of Solidarity

The SDGs depend on the idea that worldwide worries require cooperative endeavors. No single country or element can address the intricate trap of hardships we stand up to the present time. The force of unification isn't restricted to legislatures and worldwide associations; it likewise stretches out to enterprises, colleges, common society, and people. A source of inspiration addresses both the youthful dissident calling for environmental change and the corporate leader supporting maintainable strategic policies. The SDGs advise us that everybody plays a part to play in forming our reality.

Abandoning Nobody

One of the most astonishing qualities of the SDGs is its unflinching obligation to abandoning nobody. This promise to inclusivity accentuates the ethical commitment to help the oppressed and powerless. In reality as we know it where disparity proceeds, the SDGs challenge us to address the requirements of the people who are habitually overlooked or minimized. The SDGs contend for a general public where each individual's respect and privileges are secured, whether through neediness destruction, admittance to medical services, or guaranteeing clean water and disinfection.

A Strategy

The SDGs are something beyond elevated objectives; they offer a guide for activity. Legislatures and establishments all over the planet are changing these points into unmistakable approaches and exercises. Nations are realigning their public objectives to match with the SDGs, understanding that the achievement of one objective regularly reinforces the advancement of others. Creative coordinated efforts are developing among states, industry, and common society to amplify the effect of assets and abilities. The plan is being woven into the texture of worldwide administration, driving arrangement choices and venture drives.

The Effect of Development

In this time of quick mechanical development, development shows up as an impressive partner in the accomplishment of the SDGs. From sustainable power arrangements that battle environmental change to computerized stages that further develop training availability, advancement builds our capability to speed up progress. The SDGs call for using the force of advancement not simply to address challenges, but rather additionally to jump old formative ways. Development permits us to reevaluate obsolete ideal models and clear the way for additional opportunities.

Neighborhood Activity with Worldwide Effect

While the SDGs are a worldwide system, their effect is seen most capably at the neighborhood level. Networks all through the world are embracing the points and changing them to their particular settings. Nearby exercises, whether they are a grassroots mission to advance feasible farming or a local venture to guarantee clean water supplies, are building blocks that all in all make a worldwide structure of progress. The SDGs empower people to become specialists of change in their nearby networks, adding to the more extensive embroidery of progress.

Instruction and Mindfulness as Impetuses

Understanding the SDGs requires more than basically administrative changes; it requires a change in thinking and lead. Training and mindfulness are the main thrusts behind this pattern. People feel engaged to pursue reasonable decisions in their day to day routines as they become more mindful of the objectives and their suggestions. Youngsters find out about natural insurance and civil rights, ingraining a feeling of obligation for the world they will acquire. The far reaching influence of schooling and mindfulness spreads through families, organizations, and social orders, giving a cooperative force toward accomplishing the SDGs.

Challenges on the Way Forward

While the SDGs motivate trust, they likewise compel us to defy the issues that lie ahead. Complex worries, for example, environmental change, shamefulness, and savagery are profoundly settled and impervious to speedy cures. The SDGs require a change in perspective by they way we tackle these troubles. It requires a shift away from transient reasoning and toward long haul manageability. It involves building associations that cross international limits. Furthermore, it drives us to break down and adjust our system despite changing worldwide elements continually.

The Significance of the Present

As the globe wrestles with emergencies and vulnerability, the significance of accomplishing the SDGs becomes more clear. The clock is ticking, and the cutoff time of 2030 is drawing closer. Each day that passes is a botched an open door to make a superior future for people in the future. The SDGs advise us that the second for activity is currently and that our consolidated will and determine can beat even the most serious issues. The plan for worldwide activity is before us, ready to be transformed into significant change.

A Tradition of Probability

The SDGs' heritage is one of probability — a declaration to what humankind can accomplish when joined by a solitary objective. They act as a wake up call that, while our issues are tremendous and complex, our ability for resourcefulness, sympathy, and cooperation is considerably more noteworthy. The SDGs rise above political convictions and social limits, joining us under the banner of shared mankind. They show a course to a world in which all people can succeed, equity rules, and the planet is supported for people in the future.

The Unified Countries Feasible Improvement Objectives are an entrancing part in the embroidery of mankind's set of experiences, a recognition for our longing for a superior future. As we push toward 2030 and then some, let us regard the call of the SDGs, which are more than a guide; they are a promise to build a future that mirrors our most elevated desires and most prominent potential.

The Change Expanding influence

The excellence of the Reasonable Improvement Objectives rests in their elevated goals as well as in their capacity to create a wave result of progress. At the point when one objective is sought after, it often begins a chain response that benefits different objectives. Putting resources into incredible instruction, for instance, furnishes people with data and abilities, yet in addition adds to further developed wellbeing, diminished disparities, and more monetary possibilities. Essentially, progresses in clean energy advancements address environmental change as well as lessen neediness by offering admittance to reasonable and economical energy sources.

These joined advancement courses amplify the effect of our endeavors. As legislatures, associations, and people interface their activities with the SDGs, they unexpectedly add to a more all encompassing and incorporated way to deal with worldwide turn of events. This expanding influence extends across borders, reverberating through all areas and networks, delineating the tremendous capability of deliberate activity toward a typical vision.

The Business Case for Maintainability

The confidential area is basic to accomplishing the SDGs. Organizations have perceived that economical practices are morally dependable as well as decisively sound. Corporate social obligation has developed into a promise to make shared incentive for both the organization and society. Organizations that integrate supportability into their basic activities are stronger, imaginative, and cutthroat over the long haul. This financial need has brought forth a huge number of drives going from sustainable power speculations to round economy models that diminish squander and expand asset productivity.

The SDGs furnish organizations with a structure for adjusting their goals to cultural requests. As the driving force of monetary development, the confidential area has the ability to impact significant change by using its assets, information, and overall reach. Organizations can be planners of a fruitful and fair future by taking on manageable practices and adding to the fulfillment of the SDGs.

Enabling Youth

Youthful voices resonating with desperation, enthusiasm, and resolve are among the most energetic allies of the SDGs. The present children are not content to acquire a world troubled by the missteps of the past; they are influencers, impetuses of progress. They are reclassifying the account of what is conceivable with their unrestrained optimism and mechanical ability.

Youngsters are at the vanguard of the SDG development, putting together environment strikes and advocating advanced answers for social great. They request responsibility from state run administrations and establishments, requesting that they focus on the future over transient benefits. The young perceive that the SDGs are not dynamic thoughts, but instead a substantial guide that might lead the way for the general public they envision. Their energy and backing revive the overall battle for reasonable turn of events.

Organizations Have Extraordinary Power

Understanding the SDGs need a shift away from divided techniques and toward imaginative organizations. No single body holds the assets in general and arrangements expected to address muddled circumstances. Joint efforts between legislatures, community society, the scholarly world, and organizations are consequently basic to understanding the SDGs' maximum capacity.

Organizations invigorate imagination by uniting shifted thoughts and encounters. They pool assets, share data, and duplicate impact. Multi-partner joint efforts can possibly impact fundamental change and address gives that cross public limits. These joint efforts are not restricted to specific businesses; they are joined by a common craving for a superior society. The SDGs become a live reality through aggregate exertion, instead of a theoretical idea.

Progress and Responsibility

The way to the SDGs involves something other than setting targets; it additionally involves following advancement and considering partners capable. Estimation and detailing techniques are expected to ensure that vows are converted into unmistakable outcomes. States, associations, and people should look at their commitments routinely and recognize regions for development.

Straightforward and solid information is a foundation of this responsibility. It illuminates navigation, distinguishes holes, and coordinates asset allotment. Deliberate Public Surveys, in which countries report on their advancement toward the SDGs, act as stages for trading best practices and gaining from one another's encounters. This accentuation on information driven responsibility changes the SDGs from desires to noteworthy drives.

Building Strength Notwithstanding Affliction

The world isn't static; it is dynamic and always showing signs of change. Pandemics, cataclysmic events, and international pressures can hinder progress toward the SDGs. Notwithstanding, the center of the SDGs rests in their flexibility — the capacity to adjust, recuperate, and continue to push ahead regardless of mishaps.

The Coronavirus pandemic, for instance, featured the meaning of planning strong frameworks that can endure shocks. It underscored the significance of available medical care, social wellbeing nets, and web availability. As opposed to deterring us, such issues act as spikes for development and reaffirm the significance of the SDGs. They advise us that our obligation to maintainability is certainly not a fleeting interest; it is an unfaltering obligation to making a world that can persevere and flourish despite difficulty.

A Reimagined Future

As we enter the twenty-first hundred years, the Unified Countries Reasonable Improvement Objectives give a manual for exploring the intricacy of our day. They show our ability for empathy, joint effort, and change. The SDGs stress that the issues we go up against are not impossible; rather, they are amazing open doors for aggregate activity and worldwide participation.

The guide for worldwide activity is a unique report that is continually modified by the choices we make today. It welcomes us to ascend past our disparities and work toward a common future in which flourishing is shared, equity rules, and the earth flourishes. The excursion toward the SDGs is something other than a journey of improvement; it is a journey of trust — a journey that, when embraced all in all, can possibly reshape the direction of our planet.

Recover

7.2 Environment Arrangements:

Cooperative Endeavors to Battle Environmental Change

Even with the existential danger presented by environmental change, the globe has arrived at a surprising place of understanding: no country remains solitary in this conflict. An unnatural weather change, softening ice covers, and the acceleration of cataclysmic events are difficulties that rise above boundaries and belief systems. Environmental change requires a coordinated reaction, a musical orchestra of activity arranged by global joint effort and official arrangements. Enter the phase of environment arrangements, a recognition for mankind's capacity to team up for the protection of our planet and the ages who call it home

A Worldwide Orchestra of Responsibility

Environment arrangements address the world's planned work to address the basic requirement for environment activity. They represent an affirmation that the repercussions of environmental change are not restricted to geographic limits; they resonate across economies, environments, and society. These agreements are something other than bits of paper; they are promises that interface countries to a common obligation of safeguarding the climate for current and people in the future.

From the Kyoto Convention to the Paris Arrangement, these deals have filled in as turning points in the development of worldwide environment administration. They address a mindfulness that, while the causes and effects of environmental change fluctuate, arrangements require a cooperative exertion. Environment arrangements organize a worldwide orchestra of responsibility where every country assumes an unmistakable part while adding to the agreeable journey for a maintainable future.

The Paris Arrangement: A Hint of something to look forward to

The Paris Understanding, which has been proclaimed as a turning point in environment discretion, shows the force of cooperation. This noteworthy understanding, endorsed in 2015, unites essentially all nations with a common objective: to keep a dangerous atmospheric deviation well under 2 degrees Celsius above pre-modern levels, with endeavors to keep it beneath 1.5 degrees Celsius. The understanding not just perceives the direness of the environment calamity yet in addition the unbalanced effect of environmental change on weak gatherings.

The granular perspective of the Paris Understanding makes it so captivating. Rather than forcing uniform targets, it allows every country to characterize its own broadly resolved commitments (NDCs) in view of its specific conditions and abilities. This versatile construction empowers inclusion and proprietorship, laying out a feeling of shared liability while protecting public power. The Paris Understanding stands as a light of positive thinking, a demonstration of the world's capacity to meet up for everyone's benefit.

From Manner of speaking to Activity

Environment arrangements are not only political discussions; they are systems that drive substantial activity. They spur nations to carry out approaches that support sustainable power, cut emanations, and reinforce flexibility. These agreements act as impetuses for advancement, empowering states to put resources into practical innovation, move to green economies, and focus on environment transformation.

Besides, environment arrangements advance worldwide cooperation by empowering mechanical exchanges and monetary help to emerging nations. They recognize that fighting environmental change is a cooperative exertion, particularly for nations with little assets to focus on maintainability exercises. Environment reserves, for example, the Green Environment Asset, advance the progression of monetary assets to help variation and alleviation projects in weak regions, overcoming any barrier among way of talking and activity.

Past Legislatures: The Job of Non-State Entertainers

Environment arrangements include legislatures as well as a different scope of non-state entertainers. Organizations, towns, common society associations, and individuals have made that big appearance as dynamic entertainers in the battle against environmental change. The "We Are Still In" drive, for instance, saw a huge number of towns, enterprises, colleges, and states in the US promise to safeguard the upsides of the Paris Understanding even after the nation chose to pull out.

The cooperation of non-state entertainers amplifies the effect of environment arrangements. It demonstrates that environment activity isn't restricted to the passageways of force; it reverberates with individuals from varying backgrounds. Organizations focus on carbon lack of bias, states embrace manageable metropolitan preparation, and people settle on informed decisions that diminish their carbon impression. This collective exertion resounds, making a flood of energy that upholds global responsibilities.

The Street Ahead: Difficulties and Potential open doors

While environment arrangements address trust, they likewise uncover the obstructions that lie ahead. The earnestness of the environment circumstance requires prompt activity, and the ongoing pace of progress misses the mark regarding the objectives set by these arrangements. The pressure between financial extension and ecological assurance perseveres, requiring novel methods that balance the two objectives.

In any case, issues habitually make opportunities for change. The shift to a low-carbon economy can possibly make green positions, advance mechanical development, and change businesses. Environment arrangements can go about as prods for strategy advancement, empowering states to embrace powerful environment systems that encourage reasonable turn of events.

A Common Predetermination

Environment arrangements are something beyond conciliatory amenities; they represent our normal fate.

They embody the conviction that our aggregate will is more grounded than the powers that jeopardize our reality. As we work to restrict a dangerous atmospheric deviation, defend biodiversity, and establish environment strength, these arrangements act as beacons driving us across the violent oceans of environmental change.

The ensemble of joint effort keeps on playing, reverberating through discussions, strategy, and individual deeds. Each vow made every fossil fuel byproduct cut, adds to the crescendo of trust — trust that rises above boundaries, societies, and contrasts. Environment arrangements advise us that, while the difficulties are perfect, our capacity for change is as well. We can change the tale of our planet's future together, transforming it into one of flexibility, stewardship, and a common obligation to a reasonable world.

7.3 Transboundary Protection:

Safeguarding Shared Biological systems and Assets

Transboundary Protection: Safeguarding Shared Environments and Assets, Transboundary protection arises as a light of expectation for preserving our planet's most important biological systems in this present reality where nature knows no boundaries. The world's biodiversity doesn't regard political limits; it streams openly, connecting changed conditions and species in perplexing ways. Transboundary preservation perceives this interconnectivity by crossing international limits to make an orchestra of insurance that reverberations across scenes, species, and ages.

Biodiversity Crossing Limits

Transboundary protection shows humankind's capacity to team up even with environmental worries. It perceives that waterways, mountains, and woods don't stop at imaginary lines painted on maps. All things being equal, they wander and extend, producing natural connections that rise above public limits. These binds need an all encompassing way to deal with protection, one that arranges adjoining nations as guardians of shared assets. Transboundary protection is certainly not a decision however a need, from the fabulous African savannas that range various nations to the huge Amazon rainforest that covers different South American nations. It figures out that the endurance of explicit species, the protection of delicate environments, and the accessibility of significant assets are reliant upon serene cooperation among states.

The Coordinated effort Impact

The gradually expanding influence of coordinated effort is vital to transboundary preservation. At the point when legislatures rally to protect shared biological systems, the effect stretches out past individual limits. Endeavors to control unlawful natural life exchange, for instance, become additional compelling when policing from adjoining nations team up. Reclamation projects, for example, transboundary stream restoration, build up some momentum when nations pool their assets and aptitude.

The far reaching influence spreads to the networks that live in or close these transboundary areas. Protection projects that foster cradle zones and reasonable administration rehearses habitually work on the vocations of neighborhood occupants. Ecotourism projects, logical exploration, and social trades advance monetary development and improve the texture of these networks. Transboundary protection in this manner fills in as a driver for bigger provincial turn of events.

Conservation and the Diplomatic Spirit

Transboundary preservation requires natural information as well as the soul of discretion. Arranging accords and structures that suit the interests of numerous countries while safeguarding environmental uprightness requires strategic abilities. It requires a common vision, legit discussion, and an eagerness to think twice about.

The Global Association for Protection of Nature (IUCN) perceives the worth of transboundary coordinated effort through its Transboundary Preservation Expert Gathering.

This gathering gives best practices direction and a setting for nations to profit from each other's encounters. Such joint exercises help in the interpretation of transboundary preservation thoughts into real measures.

Past Nature: Social and Authentic Significance

Transboundary preservation contains something beyond environmental significance; it likewise incorporates social and authentic importance. A large number of these connected environments have extraordinary social and otherworldly importance for native gatherings. Protecting these regions is about something beyond safeguarding biodiversity; additionally about safeguarding legacy and customs have existed for ages.

Also, transboundary protection much of the time covers with authentic stories. Due to its restricted admittance, the neutral ground (DMZ) among North and South Korea, for instance, has accidentally turned into a sanctuary for creatures. These unmistakable situations feature the convoluted connection between political history, human exercises, and natural outcomes.

Wins and Difficulties

While the idea of transboundary preservation is moving, it isn't without impediments. Different guidelines, contending interests, and authentic strains can all hamper cooperative endeavors. In any case, triumphs in this field act as tokens of the capability of aggregate activity.

One renowned model is the Harmony Parks Establishment, which spearheaded the foundation of transboundary preservation saves in Southern Africa. These parks, which range various nations, have saved natural life as well as advanced tact and local area improvement. They are substantial evidence that protection can prompt conciliatory leap forwards and further developed ways of life.

Transboundary Trust

Transboundary preservation arises as a light of trust in this present reality where natural difficulties have no limits. It communicates the acknowledgment that the fate of our planet's most significant biological systems is in our aggregate hands. This cooperative methodology perceives the profound linkages that string across nature, as well as the way that rationing shared assets requires shared liability.

Transboundary protection is a statement that we can ascend past divisions to support the planet's wellbeing and the prosperity of its occupants. It fills in as an update that our exercises affect far off districts and people in the future. By tolerating transboundary preservation standards, we rise above public limits as well as the hole between human desire and the obligation to safeguard the regular world.

An Embroidery of Solidarity: Transboundary Preservation in real life

The narratives of transboundary protection exercises weave an embroidery of harmony that traverses mainlands and biological systems. The Selous-Niassa Untamed life Passageway, which interfaces Tanzania's Selous Game Save and Mozambique's Niassa Game Hold, is one such stupendous model. This hall gives a protected section to elephants and different species, diminishing environment fracture and guaranteeing hereditary assortment among populaces. By crossing borders, this work shows how collaboration might carry shared dreams of natural insurance to the real world.

The European Green Belt is one more illustration of the strength of transboundary protection. This organization of safeguarded spots and natural surroundings extends north of 12,500 kilometers and crosses 24 countries, following the way of the previous Iron Shade. Nature has transformed what was once an indication of division into an image of solidarity. The European Green Belt energizes cross-line joint effort, environment reclamation, and the renewed introduction of compromised species populaces.

A Green Extension to Transboundary Harmony

The connection between transboundary preservation and harmony is perplexing and critical. Shared environments regularly rise above verifiable clash zones, offering a common stage for joint effort. The Wadden Ocean, an UNESCO World Legacy site, extends along the shores of Denmark, Germany, and the Netherlands. Security is a binding together movement rises above verifiable contention, featuring the normal obligation of moderating this vital territory for relocating birds and marine life.

The Brilliant Entryway Harmony Park is an image of mending and cooperation between South Korea and China. This transboundary preservation region traverses their boundaries, advancing biological reclamation and discretionary commitment. What was once a wellspring of conflict is presently an image of idealism, showing that nature can act as a scaffold to understanding and joint effort.

Association Innovation

Transboundary preservation has acquired another partner in the period of present day innovation. Remote detecting, satellite imaging, and geographic data frameworks empower ongoing observing and information sharing across borders. This mechanical interconnection benefits in following natural life relocations, distinguishing unlawful movement, and surveying ecological changes. Countries can team up to oversee and investigate ecological information across borders utilizing stages like the Natural Frameworks Exploration Organization (ESRI).

Nearby Commitment: The Soul of Transboundary Achievement

The outcome of transboundary preservation is reliant on peaceful accords, yet in addition on neighborhood commitment. The investment of native gatherings, neighborhood residents, and partners is basic for long haul maintainability. These people group regularly have a significant familiarity with their biological systems, and their support guarantees that protection endeavors are socially conscious as well as financially practical.

Transboundary preservation programs flourish when they fit with the objectives of these populaces. At the point when neighborhood occupants see the advantages of preservation, like better occupations, more noteworthy eco-the travel industry, and reinforced social linkages, they become dynamic associates in safeguarding shared assets. This grassroots commitment lays areas of strength for joint protection endeavors to flourish.

A Shared Legacy for Future Generations

Transboundary preservation is an interest in the heritage we leave for people in the future. It exhibits our capacity to transcend conflicts and worth the prosperity of the world over transient benefits. Achievement stories of cross-line safeguarded zones are encouraging signs that enlighten the way forward. As we face the challenges of an evolving environment, declining biodiversity, and modifying biological systems, the ideas of transboundary preservation become progressively significant. They advise us that our worldwide local area is connected in terms of professional career and innovation as well as by the delicate strings of nature that support every one of us. We can wind around a powerful preservation texture that will keep going long into the future by means of shared vision, cooperation, and commitment.

7.4 Diverse Maintainability: Gaining from Different Methodologies, In a world sew together by strings of variety, the objective of maintainability takes on extra aspects when analyzed through the crystal of multifaceted comprehension. The criticalness of tending to worldwide worries, for example, environmental change and asset consumption, ties mankind in a typical drive for a superior future. Notwithstanding, the ways we take to accomplish supportability are shaped by our singular accounts, customs, and points of view. Diverse maintainability praises this complicated embroidery of points of view and requests that we acknowledge the insight that emerges when many voices join on a comparative goal.

A Worldwide Points of view Mosaic

View yourself as in a packed market where many societies impact. Each slow down offers an investigate an alternate culture, with shifted cooking styles, splendid materials, and music that tell the recollections of previous eras. Likewise, the universe of supportability is an energetic commercial center of thoughts, with each culture bringing an interesting perspective that works on the aggregate excursion.

Native social orders, for instance, furnish experiences into the agreeable conjunction with a climate that has kept up with them for a long time. Their broad information on biological systems and interconnectedness with the earth could rouse current supportability strategies in view of regard and correspondence. Likewise, Eastern customs stress equilibrium and network, advising us that the journey for progression doesn't need to be withdrawn from profound prosperity.

Making a Worldwide Discourse Diverse maintainability lights a worldwide conversation that rises above topographical limits. It's a challenge to go along with us at a table where stories, practices, and thoughts are shared. At the point when Maasai pastoralists from Africa share their practices for feasible land the executives, and Japanese people group present thoughts like "mottainai" (a feeling of regret over squander), a rich progression of thoughts results. This talk assists us with adjusting and integrating rehearses that are pertinent to our particular settings, while likewise developing a shared mindset of our common planet.

Custom's Insight, Our progenitors lived economically out of need, adjusting their lives to the rhythms of the normal world. Native societies, specifically, incorporate an abundance of information that can be applied to current supportability endeavors. From the Navajo thought of "hosho" (a condition of equilibrium and concordance) to the Andean way of thinking of "ayni" (correspondence with nature), these customary qualities reverberate the rules that drive a supportable world. By perceiving and regarding these practices, we gain admittance to an abundance of information that might impact our choices and shape our strategies. Their lessons advise us that maintainability isn't a trend, however an old truth that has endured everyday hardship.

Rising above Straight Movement, The Western fantasy of progress every now and again interfaces improvement with straight development and gathering. Diverse maintainability, then again, urges us to reevaluate this story. Native perspectives, for instance, advocate for repeating ideas of time, in which each activity resounds in patterns of circumstances and logical results. This perspective urges us to embrace round economies, in which assets are esteemed, reused and recovered as opposed to removed and disposed of.

Integrating old recurrent types of thought into present day structures can possibly reform how we approach utilization, squander the executives, and financial development. It is a progress from the quest for vast extension to the development of long haul balance.

The Impact of Limitation, Multifaceted supportability additionally accentuates the significance of limited arrangements. While worldwide objectives give a widespread structure, their execution ought to be redone to the singular states of networks. What works in a provincial local area in the Amazon rainforest may not be appropriate for an Asian uber city. Embracing different viewpoints empowers us to foster special arrangements that emerge from the grassroots, utilizing nearby information and assets.

This restriction guarantees the significance of maintainability programs as well as fosters a feeling of responsibility and strengthening inside networks. It accentuates that the best arrangements much of the time come from inside, driven by the profound association among individuals and their settings.

Variety as an impetus for development Inventiveness creates when developments impact. Culturally diverse manageability is a blend where one of a kind arrangements rise out of the combination of thoughts. Supportable structures that can oppose outrageous weather conditions are made when Scandinavian building thoughts are joined with Center Eastern conventional plan. While antiquated cultivating strategies from Asia meet present day agroecology, versatile farming frameworks arise. the range of strategies advances novel thoughts, empowering us to address suppositions and adventure into a new area. The gathering of various perspectives makes a cooperative energy that can drive maintainability endeavors higher than ever.

The Call for Consideration, Multifaceted manageability advises us that the way to a more economical world should be comprehensive. It features the need to intensify underrepresented voices, perceive verifiable treacheries, and perceive the different effect of ecological worries on assorted gatherings. Comprehensive supportability looks for civil rights, value, and the strengthening, everything being equal, notwithstanding natural points.

We stay away from a one-size-fits-all methodology by inviting various suppositions, which could unexpectedly sustain imbalance. All things considered, we outline a course in which each culture adds to a mosaic of arrangements that on the whole advantage humankind.

An Agreeable Ensemble of Progress

Multifaceted manageability imagines a delightful orchestra of progress wherein the insight of ages and the developments of today consolidate without a hitch. It perceives that manageability is a diverse precious stone that sparkles most brilliantly when seen from the perspective of fluctuated sees.

As we arrange the convoluted trap of worldwide troubles, remember that the arrangements we look for might be nearer than we understand, resting in the hearts and customs of developments the whole way across the world. Diverse manageability provokes us to remain at the junction of assortment, where grasping, joint effort, and normal reason prepare to a more promising time to come.

CONCLUSION

Embracing Supportability as a Lifestyle is presently not simply a decision; it is a convincing prerequisite in this present reality where the earnest interest for natural protection is turning out to be increasingly loud as time passes. Manageability is a crucial mindset shift and a progressive approach to living that holds the way into a superior future for our planet and its kin. It isn't simply confined to eco-accommodating things or stylish popular expressions. Envision residing in a general public where each individual's choices, regardless of how immaterial, have a positive far reaching influence. Settling on purposeful choices that fit with nature's rhythms is fundamental to embracing manageability as a lifestyle. Understanding the delicate harmony between human existence and the climate and acting proactively to fix the harm we've done is critical. This thought goes past reusing and utilizing energy-saving lights; about a thorough change influences how we live, eat, travel, consume, and draw in with our current circumstance.

Embracing supportability on a very basic level involves reevaluating our relationship with assets. It includes picking things with a little natural engraving and helping organizations that put a high worth on moral way of behaving. Feasible choices are driving the way for organizations that regard the climate and the freedoms of representatives in various areas, including style and food. We send areas of strength for a that we favor higher standards without ever compromising and manageability over superfluity by supporting these endeavors.

The center of taking on maintainability as a lifestyle dwells in mindfulness. It's tied in with being intensely aware of the impacts of our way of behaving and understanding that the world our youngsters acquire tomorrow is molded by the choices we make now. Each decision matters, whether it's restricting the utilization of single-use plastics, saving water, or selecting to walk as opposed to drive. By playing out these apparently little deeds, we diminish our carbon impression as well as spur individuals around us to do likewise.

 The expanding influence of maintainability is one astounding component. We become change specialists in our networks when we embrace maintainable ways of behaving. Individuals discuss our choices, which motivates them to rethink their schedules and inspect various choices. We fabricate a chain response through these conversations that could bring about bunch activity. At the point when individuals rally under the standard of manageability, the conceivable outcomes range from neighborhood tidy up missions to local area gardens.

Taking on maintainability as a lifestyle doesn't need surrendering comfort or solace. It includes thinking of inventive responses that accommodate advancement with protection. The globe is spilling over with splendid cerebrums chipping away at developments in squander decrease, sustainable power, and regenerative horticulture. We can advance and take on these developments and assist with making a maintainable future without forfeiting the delights of current living.

Taking on manageability is, at its center, an assertion of hopefulness. Indeed, even despite overpowering natural worries, it is a choice to put stock in the chance of good change. It takes us all cooperating to moderate the planet's numerous environments, safeguard its regular wonders, and ensure that people in the future might partake in its quality. Meshing supportability into our day to day existence makes us ready for a more quiet dwelling together with the regular world.

As we pick supportability as a lifestyle, we assume the stewardship of the planet and make an account of consideration, responsibility, and recovery. Each eco-accommodating decision, a significant investment of time and energy to lessen squander, and each asset saving activity adds to this continuous story. So we should execute these actions all in all, with courageous determination, and impact the world with each smart activity in turn. Manageability, all things considered, isn't just an objective; an excursion with huge importance begins with the choices we make.